T0212223

Models of Horizontal Eye Movements

Part 4

A Multiscale Neuron and Muscle Fiber-Based Linear Saccade Model

Synthesis Lectures on Biomedical Engineering

Editor
John D. Enderle, *University of Connecticut*

Lectures in Biomedical Engineering will be comprised of 75- to 150-page publications on advanced and state-of-the-art topics that span the field of biomedical engineering, from the atom and molecule to large diagnostic equipment. Each lecture covers, for that topic, the fundamental principles in a unified manner, develops underlying concepts needed for sequential material, and progresses to more advanced topics. Computer software and multimedia, when appropriate and available, are included for simulation, computation, visualization and design. The authors selected to write the lectures are leading experts on the subject who have extensive background in theory, application and design. The series is designed to meet the demands of the 21st century technology and the rapid advancements in the all-encompassing field of biomedical engineering that includes biochemical processes, biomaterials, biomechanics, bioinstrumentation, physiological modeling, biosignal processing, bioinformatics, biocomplexity, medical and molecular imaging, rehabilitation engineering, biomimetic nano-electrokinetics, biosensors, biotechnology, clinical engineering, biomedical devices, drug discovery and delivery systems, tissue engineering, proteomics, functional genomics, and molecular and cellular engineering.

Models of Horizontal Eye Movements: Part 4, A *Multiscale* Neuron and Muscle Fiber-Based Linear Saccade Model
Alireza Ghahari and John D. Enderle
2015

Mechanical Testing for the Biomechanics Engineer: A Practical Guide
Marnie M. Saunders
2015

Models of Horizontal Eye Movements: Part 3, A Neuron and Muscle Based Linear Saccade Model
Alireza Ghahari and John D. Enderle
2014

Digital Image Processing for Ophthalmology: Detection and Modeling of Retinal Vascular Architecture
Faraz Oloumi, Rangaraj M. Rangayyan, and Anna L. Ells
2014

© Springer Nature Switzerland AG 2022
Reprint of original edition © Morgan & Claypool 2015

All rights reserved. No part of this publication may be reproduced, stored in a retrieval system, or transmitted in any form or by any means—electronic, mechanical, photocopy, recording, or any other except for brief quotations in printed reviews, without the prior permission of the publisher.

Models of Horizontal Eye Movements:
Part 4, A *Multiscale* Neuron and Muscle Fiber-Based Linear Saccade Model

Alireza Ghahari and John D. Enderle

ISBN: 978-3-031-00535-0   paperback
ISBN: 978-3-031-01663-9   ebook

DOI 10.1007/978-3-031-01663-9

A Publication in the Springer series
SYNTHESIS LECTURES ON BIOMEDICAL ENGINEERING

Lecture #55
Series Editor: John D. Enderle, *University of Connecticut*
Series ISSN
Print 1930-0328   Electronic 1930-0336

Models of
Horizontal Eye Movements

Part 4

A *Multiscale* Neuron and Muscle Fiber-Based Linear Saccade Model

Alireza Ghahari
University of Connecticut

John D. Enderle
University of Connecticut

SYNTHESIS LECTURES ON BIOMEDICAL ENGINEERING #55

ABSTRACT

There are five different types of eye movements: saccades, smooth pursuit, vestibular ocular eye movements, optokinetic eye movements, and vergence eye movements. The purpose of this book series is focused primarily on mathematical models of the horizontal saccadic eye movement system and the smooth pursuit system, rather than on how visual information is processed.

In Part 1, early models of saccades and smooth pursuit are presented. A number of oculomotor plant models are described here beginning with the Westheimer model published in 1954, and up through our 1995 model involving a 4th order oculomotor plant model. In Part 2, a 2009 version of a state-of-the-art model is presented for horizontal saccades that is 3rd-order and linear, and controlled by a physiologically based time-optimal neural network. Part 3 describes a model of the saccade system, focusing on the neural network. It presents a neural network model of biophysical neurons in the midbrain for controlling oculomotor muscles during horizontal human saccades.

In this book, a multiscale model of the saccade system is presented, focusing on a multiscale neural network and muscle fiber model. Chapter 1 presents a comprehensive model for the control of horizontal saccades using a muscle fiber model for the lateral and medial rectus muscles. The importance of this model is that each muscle fiber has a separate neural input. This model is robust and accounts for the neural activity for both large and small saccades. The muscle fiber model consists of serial sequences of muscle fibers in parallel with other serial sequences of muscle fibers. Each muscle fiber is described by a parallel combination of a linear length tension element, viscous element, and active-state tension generator. Chapter 2 presents a biophysically realistic neural network model in the midbrain to drive a muscle fiber oculomotor plant during horizontal monkey saccades. Neural circuitry, including omnipause neuron, premotor excitatory and inhibitory burst neurons, long lead burst neuron, tonic neuron, interneuron, abducens nucleus, and oculomotor nucleus, is developed to examine saccade dynamics. The time-optimal control mechanism demonstrates how the neural commands are encoded in the downstream saccadic pathway by realization of agonist and antagonist controller models. Consequently, each agonist muscle fiber is stimulated by an agonist neuron, while an antagonist muscle fiber is unstimulated by a pause and step from the antagonist neuron. It is concluded that the neural network is constrained by a minimum duration of the agonist pulse, and that the most dominant factor in determining the saccade magnitude is the number of active neurons for the small saccades. For the large saccades, however, the duration of agonist burst firing significantly affects the control of saccades. The proposed saccadic circuitry establishes a complete model of saccade generation since it not only includes the neural circuits at both the premotor and motor stages of the saccade generator, but it also uses a time-optimal controller to yield the desired saccade magnitude.

KEYWORDS

burst firing, compartmental approach, muscle fiber, neural dynamic, neural input, neural model, neural network, oculomotor plant, saccade, system identification, time-optimal control

Contents

Acknowledgments

We wish to express our thanks to William Pruehsner for drawing many of the illustrations in this book.

Alireza Ghahari and John D. Enderle
January 2015

CHAPTER 1

A New Linear Muscle Fiber Model for Neural Control of Saccades[1]

1.1 INTRODUCTION

A fast eye movement is usually referred to as a saccade, and involves quickly moving the eye from one image to another image. This type of eye movement is very common, and it is observed most easily while reading—that is, when the end of a line is reached, the eyes are moved quickly to the beginning of the next line.

This chapter updates a neural network that controls the eyes during horizontal saccades and introduces a new muscle fiber model. The previous published model uses an anatomically and physiologically correct model of the oculomotor plant and neural network [Enderle and Zhou, 2010, Zhou et al., 2009]. A key element of the neural network involves the autonomous burst firing of the excitatory burst neuron (EBN) [Zhou et al., 2009] and post inhibitory rebound burst (PIRB) firing in the paramedian pontine reticular formation. In that study, the neural firing rate in for the agonist and antagonist motoneurons were separately estimated using the system identification technique. Here, each muscle fiber has a separate neural input that allows a more precise control of the saccade, which allows the investigation of some shortcomings of the previous model.

The time-optimal controller described by Enderle and Zhou [2010] has a firing rate in *individual* neurons that is maximal during the agonist pulse and independent of eye orientation, while the antagonist muscle is inhibited. In this model, the activity of all neurons is summarized into the firing of a single neuron. Thus, as the magnitude of the saccades increases, the firing rate of the single neuron increases up to $8°$, after which the neuron fires maximally. However, the firing rate of a real neuron is maximal and does not change as a function of saccade magnitude. To explain this, the time-optimal controller was hypothesized to operate in two modes—one for small saccades and one for large saccades—based on the number of neurons activated by the

---

[1]Some of the material in this chapter is an expansion of a previously published paper: Enderle, J. and Sierra, D. (2013) A new linear muscle fiber model for neural control of saccades. Int. J. of Neural Systems, 23(2). DOI: 10.1142/S0129065713500020. This paper was selected as the *2013 Hojjat Adeli Award for Outstanding Contribution in Neural Systems*. The *Hojjat Adeli Award for Outstanding Contributions in Neural Systems*, established by World Scientific Publishing Co. in 2010, is awarded annually to the most innovative paper published in the previous volume/year of the International Journal of Neural Systems.

Superior Colliculus. The subject of this chapter is to introduce a muscle fiber model of muscle, which, when incorporated into the oculomotor plant, allows the use of multiple neurons to drive the eyes to their destination.

Zhou and coworkers presented a linear third-order model of the oculomotor plant for horizontal saccadic eye movements using a lumped parameter muscle model [Zhou et al., 2009]. The muscle model used in the oculomotor plant shown in Fig. 1.1, first published by Enderle and coworkers [Enderle et al., 1991], consists of a Voigt element (viscosity and elasticity elements in parallel) in series with another Voigt element in parallel with an active-state tension generator. We refer to this model as the whole muscle model in this chapter. The tension created by the muscle is T, and the variable x_i is the change from equilibrium for each node. This linear muscle model has been shown to exhibit accurate nonlinear force-velocity and length-tension relationships for the medial and lateral rectus muscles.

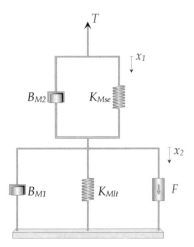

Figure 1.1: Diagram illustrates the linear muscle model consisting of an active-state tension generator F in parallel with a length-tension elastic element K_{Mlt} and viscous element B_{M1}, connected to a series elastic element K_{Mse} in parallel with a viscous element B_{M2}. Upon stimulation of the active-state tension generator F, a tension T is exerted by the muscle. We refer to this model as the whole muscle model. (Adapted from Enderle et al. [1991]).

To accurately describe the neural input to the muscles for small saccades, it is necessary to model the muscle at the basic building block level of the muscle fiber. Models of muscle typically used in the oculomotor system during saccades are lumped parameter models, that is, information about muscle fibers and other features is reduced to a small set of parameters. Further, the single neural input to the lumped parameter muscle model captures the entire population of neurons that fire and innervate the muscle. With a muscle fiber model, each muscle fiber has its own neural input allowing the impact of the number of actively firing neurons to be investigated. As demonstrated, the number of neurons firing significantly affects the control of saccades rather than variations in the firing rate among neurons.

Muscles are actuators that perform different tasks controlled by the central nervous system. Illustrated in Fig. 1.2 is the anatomy of a muscle that consists of two tendons, and a serial and parallel network of muscle fibers. The muscle fiber (an individual cell) is the smallest independent muscle unit that displays the same mechanical properties as the whole muscle. Models explicitly defined to study coordination and force generation during motor tasks include mathematical descriptions of the muscle behavior that range from the microscopic properties of the muscle to the analysis of their input-output characteristics [Zajac, 1989].

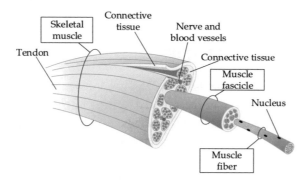

Figure 1.2: Anatomy of a muscle that consists of two tendons, and a serial and parallel network of muscle fibers.

Goldberg and coworkers [1997] report that cat lateral rectus muscle contains approximately 15,000 muscle fibers and that it is innervated by 1,100 motoneurons, which means that the average motor unit consists of 15 muscle fibers and one motoneuron. Rather than muscle fibers arranged in columns of series linked muscle fibers as describe here, the cat has muscle fibers whose architecture involves serially arranged and branching networks.

There are 20,000 to 30,000 muscle fibers in the human lateral and medial rectus muscle [Carpenter, 1988]. Leigh and Zee [1999] report that the medial and lateral rectus muscle differ anatomically, physiologically, and immunologically from skeletal muscle. Some differences include muscle fibers that are smaller and more richly innervated. These muscles, consisting of six different muscle fiber types, are the fastest contracting muscles in the body and are fatigue resistant. There are two different layers in the medial and lateral rectus muscle consisting of a central global layer and a peripheral orbital layer, with each having different ratios of muscle fibers that can either sustain contraction or provide brief rapid contraction. 80% of the orbital layer has singly innervated and fatigue resistant muscle fibers that provide the brief rapid contraction responsible for driving the eyes to their destination during saccades—these fibers have numerous mitochondria in dense clusters that are not present in skeletal muscle. The global layer has a mixture of fibers, with singly and multiply innervated, fatigue resistant, and fatigable, and provide sustained or rapid brief contraction—these muscle fibers are thought to keep the eyes at their destination after a saccade.

A scalable muscle fiber-based muscle model is introduced here that exhibits accurate non-linear force-velocity and length-tension relationships, and is incorporated into an oculomotor plant. At the muscle fiber level as shown, one can investigate the impact of the number of actively firing neurons during saccades. This type of system was suggested by Sparks [2002] as an ideal vehicle for investigating the oculomotor system.

In this chapter, the focus is on the neural input to the orbital layer of muscle fibers, rather than focusing on the different types muscle fibers. We also investigate the synchrony of neuron firing and variations of firing frequency in the neuron population.

1.2   MUSCLE FIBER MODEL

The muscle fiber model of muscle is shown in Fig. 1.3. The tendon is described with the viscous and elastic elements, B_2 and K_{se}, at the top and bottom of the figure, and the muscle fiber is described with the active-state generator F_j^i, viscous element, B_1, and elastic element, K_{lt}, where i refers to the ith muscle fiber column and j refers to the jth series muscle fiber in column i. In Fig. 1.3, there are m muscle fibers in series with two tendon elements, in parallel with n columns of other tendons and muscle fibers. The overall tension created by the muscle is T, variable x_1 is the change from equilibrium length for the muscle in the lengthening direction, and variable x_j^i is the change from equilibrium at node j in the muscle fiber column i.

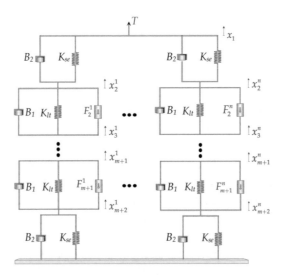

Figure 1.3: Muscle fiber model of muscle.

We have assumed that the muscle fibers are identical for simplicity, and that each muscle fiber has an active-state tension generator that can be individually stimulated using different neural inputs. In reality, extraocular muscles contain at least six different muscle fiber types that can be described as slow and fast [Sparks, 2002]. More importantly, the neural input to the muscle fiber can be appropriately selected depending on the experiment, which allows us to examine the interaction of motoneurons on the muscle (e.g., synchrony in firing, some neurons firing at different rates, and some neurons not firing), which has a profound impact on the neural control model.

The tension developed from the muscle fiber model is given by

$$
\begin{aligned}
T &= \sum_{i=1}^{n} T_i \\
&= \sum_{i=1}^{n} \left(K_{se}(x_1 - x_2^i) + B_2(\dot{x}_1 - \dot{x}_2^i) \right),
\end{aligned}
\tag{1.1}
$$

where T_i is the tension developed by each muscle fiber column. The notation $\dot{x}$ is shorthand for $\dfrac{dx}{dt}$. Using D'Alembert's principle, the equations that define each muscle fiber column are given by

$$
\begin{aligned}
T_i &= K_{se} \left(x_1 - x_2^i \right) + B_2 \left(\dot{x}_1 - \dot{x}_2^i \right), \\
K_{se} & \left(x_1 - x_2^i \right) + B_2 \left(\dot{x}_1 - \dot{x}_2^i \right) \\
&= K_{lt} \left(x_2^i - x_3^i \right) + B_1 \left(\dot{x}_2^i - \dot{x}_3^i \right) + F_2^i, \\
K_{lt} & \left(x_2^i - x_3^i \right) + B_1 \left(\dot{x}_2^i - \dot{x}_3^i \right) + F_2^i \\
&= K_{lt} \left(x_3^i - x_4^i \right) + B_1 \left(\dot{x}_3^i - \dot{x}_4^i \right) + F_3^i, \\
&\quad\vdots \\
K_{lt} & \left(x_m^i - x_{m+1}^i \right) + B_1 \left(\dot{x}_m^i - \dot{x}_{m+1}^i \right) + F_m^i \\
&= K_{lt} \left(x_{m+1}^i - x_{m+2}^i \right) + B_1 \left(\dot{x}_{m+1}^i - \dot{x}_{m+2}^i \right) + F_{m+1}^i, \\
K_{lt} & \left(x_{m+1}^i - x_{m+2}^i \right) + B_1 \left(\dot{x}_{m+1}^i - \dot{x}_{m+2}^i \right) + F_{m+1}^i \\
&= K_{se} x_{m+2}^i + B_2 \dot{x}_{m+2}^i.
\end{aligned}
\tag{1.2}
$$

To make the muscle fiber model more compact and easier to simulate, a state variable approach is used with $y_1^i = x_1 - x_2^i$, and $y_{m+2}^i = x_{m+2}^i$, and for $j = 2, \ldots, m+1$, $y_j^i = x_j^i - x_{j+1}^i$. It then follows that

$$
x_1 = \sum_{j=1}^{m+2} y_j^1 = \cdots = \sum_{j=1}^{m+2} y_j^n.
\tag{1.3}
$$

The model is now given by

$$\dot{y}_1^i = \frac{T_i - K_{se} y_1^i}{B_2},$$ (1.4)

and for $j = 2, \ldots, m + 1$,

$$\dot{y}_j^i = \frac{T_i - K_{lt} y_j^i - F_j^i}{B_1},$$ (1.5)

and

$$\dot{y}_{m+2}^i = \frac{T_i - K_{se} y_{m+2}^i}{B_2}.$$ (1.6)

1.2.1 SCALABILITY AND STEADY-STATE

The structure of the muscle fiber muscle model allows one to calculate the viscosities and elasticities as a function of the whole muscle parameter model as follows:

$$K_{se} = \frac{2 K_{Mse}}{n}, \quad K_{lt} = \frac{m K_{Mlt}}{n},$$
$$B_1 = \frac{m B_{M1}}{n}, \quad B_2 = \frac{2 B_{M2}}{n}.$$ (1.7)

Further, assuming the same active-state tension in each muscle fiber, gives the following relationship $F_j^i = \frac{F}{n}$.

To evaluate steady-state conditions for the muscle fiber muscle model, we start with Eq. (1.3) and substitute steady-state conditions from Eqs. (1.4)–(1.6) (i.e., from $\dot{y}_1^i = 0 = \frac{T_i - K_{se} y_1 (\infty)}{B_2}$, we get $y_1^i (\infty) = \frac{T_i}{K_{se}}$ and from $\dot{y}_{m+2}^i = 0 = \frac{T_i - K_{se} y_{m+2}^i (\infty)}{B_2}$, we get $y_{m+2}^i (\infty) = \frac{T_i}{K_{se}}$ and for $j = 2, \ldots, m + 1$, with $\dot{y}_j^i = 0 = \frac{T_i - K_{lt} y_j^i (\infty) - F_j^i}{B_1}$, we get $y_j^i (\infty) = \frac{T_i - F_j^i}{K_{lt}}$). The notation $y (\infty)$ refers to y at steady-state or y at $t = \infty$. Once substituted for any muscle fiber column i, gives

$$x_1 (\infty) = \sum_{j=1}^{m+2} y_j^i$$
$$= \frac{T_i (\infty)}{K_{se}} + \frac{T_i (\infty) - F_2^i}{K_{lt}} + \cdots$$
$$+ \frac{T_i (\infty) - F_{m+1}^i}{K_{lt}} + \frac{T_i (\infty)}{K_{se}}$$
$$= \left(\frac{2 K_{lt} + m K_{se}}{K_{se} K_{lt}} \right) T_i (\infty) - \frac{1}{K_{lt}} \sum_{j=2}^{m+1} F_j^i,$$ (1.8)

or in terms of tension

$$T_i(\infty) = \left(\frac{K_{se}K_{lt}}{2K_{lt} + mK_{se}} \right) \left(\frac{1}{K_{lt}} \sum_{j=2}^{m+1} F_j^i + x_1(\infty) \right). \tag{1.9}$$

Using Eqs. (1.4)–(1.6) and (1.9), the steady-state for the state variables are given by

$$\begin{aligned} y_1^i(\infty) &= \frac{T_i}{K_{se}} \\ &= \left(\frac{K_{lt}}{2K_{lt} + mK_{se}} \right) \left(\frac{1}{K_{lt}} \sum_{j=2}^{m+1} F_j^i + x_1(\infty) \right), \end{aligned} \tag{1.10}$$

and for $j = 2, \ldots, m+1$

$$\begin{aligned} y_j^i(\infty) &= \frac{T_i - F_j^i}{K_{lt}} \\ &= \left(\frac{K_{se}}{2K_{lt} + mK_{se}} \right) \left(\frac{1}{K_{lt}} \sum_{j=2}^{m+1} F_j^i + x_1(\infty) \right) - \frac{F_j^i}{K_{lt}}, \end{aligned} \tag{1.11}$$

and

$$\begin{aligned} y_{m+2}^i(\infty) &= \frac{T_i}{K_{se}} \\ &= \left(\frac{K_{lt}}{2K_{lt} + mK_{se}} \right) \left(\frac{1}{K_{lt}} \sum_{j=2}^{m+1} F_j^i + x_1(\infty) \right). \end{aligned} \tag{1.12}$$

1.2.2   STATIC AND DYNAMIC PROPERTIES OF THE MUSCLE FIBER MODEL OF MUSCLE

As shown here, the static and dynamic characteristics of the muscle fiber muscle model are identical to those of the whole muscle model demonstrated by Enderle and coworkers [1991]. It should be noted that the whole muscle model is the only linear model that has the nonlinear force-velocity and length-tension relationships observed in data.

Static Properties

To compare the length-tension characteristics between the two models, assume that all of the muscle fiber active-state tensions are identical, and then substitute the parameter values given in

Eq. (1.7) into Eq. (1.2), which gives

$$T(\infty) = \sum_{i=1}^{n} T_i(\infty)$$

$$= \sum_{i=1}^{n} \left(\frac{K_{se} K_{lt}}{2K_{lt} + mK_{se}} \right) \left(x_1(\infty) + \frac{1}{K_{lt}} \sum_{j=2}^{m+1} F_j^i \right) \Bigg|_{\substack{K_{se}=\frac{2K_{Mse}}{n} \\ K_{lt}=\frac{mK_{Mlt}}{n}}}$$

$$= \sum_{i=1}^{n} \left(\left(\frac{\dfrac{2K_{Mse}}{n} \dfrac{mK_{Mlt}}{n}}{2\dfrac{mK_{Mlt}}{n} + m\dfrac{2K_{Mse}}{n}} \right) x_1(\infty) + m \left(\frac{\dfrac{2K_{Mse}}{n}}{2\dfrac{mK_{Mlt}}{n} + m\dfrac{2K_{Mse}}{n}} \right) F_j^i \right) \tag{1.13}$$

$$= \sum_{i=1}^{n} \left(\frac{1}{n} \left(\frac{K_{Mse} K_{Mlt}}{K_{Mlt} + K_{Mse}} \right) x_1(\infty) + \left(\frac{K_{Mse}}{K_{Mlt} + K_{Mse}} \right) F_j^i \right)$$

$$= \left(\frac{K_{Mse} K_{Mlt}}{K_{Mlt} + K_{Mse}} \right) x_1(\infty) + n \left(\frac{K_{Mse}}{K_{Mlt} + K_{Mse}} \right) F_j^i \Bigg|_{F_j^i = \frac{F}{n}}$$

$$= \left(\frac{K_{Mse} K_{Mlt}}{K_{Mlt} + K_{Mse}} \right) x_1(\infty) + \left(\frac{K_{Mse}}{K_{Mlt} + K_{Mse}} \right) F. \tag{1.14}$$

Equation (1.14) is the same as Eq. (1.2) in Enderle and coworkers [1991], except that it is written terms of lengthening instead of shortening. Thus, the muscle fiber muscle model has the same length-tension characteristics, as shown in Fig. 1.3 of Enderle and coworkers [1991], which matches the data extremely well.

Dynamic Properties
To investigate the force-velocity characteristics between the two models, we once again assume that the muscle fiber active-state tensions are identical and attach a mass to the end of the muscle, which rests on a platform, as shown in Fig. 1.4. The equations that define this system are given as

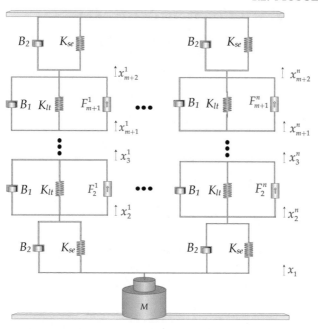

Figure 1.4: Illustration of the muscle fiber model with an attached weight resting on a platform.

$$Mg + M\ddot{x}_1 + K_{se}\left(x_1 - x_2^i\right) + B_2\left(\dot{x}_1 - \dot{x}_2^i\right) = 0,$$
$$K_{se}\left(x_1 - x_2^i\right) + B_2\left(\dot{x}_1 - \dot{x}_2^i\right)$$
$$= K_{lt}\left(x_2^i - x_3^i\right) + B_1\left(\dot{x}_2^i - \dot{x}_3^i\right) - F_2^i,$$
$$K_{lt}\left(x_2^i - x_3^i\right) + B_1\left(\dot{x}_2^i - \dot{x}_3^i\right) - F_2^i$$
$$= K_{lt}\left(x_3^i - x_4^i\right) + B_1\left(\dot{x}_3^i - \dot{x}_4^i\right) - F_3^i,$$
$$\vdots$$
$$K_{lt}\left(x_m^i - x_{m+1}^i\right) + B_1\left(\dot{x}_m^i - \dot{x}_{m+1}^i\right) - F_m^i$$
$$= K_{lt}\left(x_{m+1}^i - x_{m+2}^i\right) + B_1\left(\dot{x}_{m+1}^i - \dot{x}_{m+2}^i\right) - F_{m+1}^i,$$
$$K_{lt}(x_{m+1}^i - x_{m+2}^i) + B_1(\dot{x}_{m+1}^i - \dot{x}_m^i) - F_{m+1}^i$$
$$= K_{se}x_{m+2}^i + B_2\dot{x}_{m+2}^i.$$

$$(1.15)$$

To make the muscle fiber model more compact and easier to simulate, the state variable approach is used again with $w_1 = \dot{x}_1$, $y_1^i = x_1 - x_2^i$, and $y_{m+2}^i = x_{m+2}^i$, and for $j = 2, \ldots, m+1$, $y_j^i =$

$x_j^i - x_{j+1}^i$. With $T_i = K_{se} y_1^i + B_2 \dot{y}_1^i$, the model is given by

$$\dot{w}_1 = \frac{-(T_i + Mg)}{M}, \tag{1.16}$$

$$\dot{y}_1^i = \frac{(T_i - K_{se} y_1^i)}{B_2}, \tag{1.17}$$

and for $j = 2, \ldots, m+1$,

$$\dot{y}_j^i = \frac{T_i + F_j^i - K_{lt} y_j^i}{B_1}, \tag{1.18}$$

and

$$\dot{y}_{m+2}^i = \frac{T_i - K_{se} y_{m+2}^i}{B_2}. \tag{1.19}$$

To simulate the solution, Eqs. (1.18) and (1.19) are integrated for $i = 1, \ldots, n$, from which we compute $y_1^i = x_1 - \sum\limits_{j=2}^{m+2} y_j^i$ and $\dot{y}_1^i = \dot{x}_1 - \sum\limits_{j=2}^{m+2} \dot{y}_j^i$ for $i = 1, \ldots, n$ and then compute $T_i = K_{se} y_1^i + B_2 \dot{y}_1^i$.

Simulations for the muscle fiber model and the whole muscle model give identical results for any combination of columns of series of muscle fibers, with two examples shown in Fig. 1.5. Further, the simulation results for both muscle models and a lever system as described in Enderle and coworkers give identical results as well [1991]. The force-velocity curves in Figs. 7–9 of Enderle and coworkers are the same for either model [1991].

1.3 OCULOMOTOR PLANT

The oculomotor plant for the horizontal saccadic eye movement system is shown in Fig. 1.6. The lateral and medial rectus eye muscles are based on the muscle fiber muscle model previously described, and are stretched x_p from equilibrium at primarily position. θ is the angle the eyeball is deviated from the primary position, and variable x is the length of arc traversed. When the eye is at the primary position, both θ and x are equal to zero. The agonist muscle is on the left with overall change in length of x_{ag}, and the antagonist muscle is on the right with overall change in length of x_{ant}. It is assumed that the structure of the agonist and antagonist muscle is the same, that is, the same number of columns and sequences of muscle fibers.

Note that the passive elasticity of the eyeball, K_p, is a combination of the effects due to the four other muscles, optic nerve, etc., and is a rotational element. The viscous element of the eyeball, B_p, is due to the friction of the eyeball within the eye socket and is a rotational element. The moment of inertia of the eyeball is J_p. The radius of the eyeball is r.

The inputs in the muscle fiber model are the agonist and antagonist active-state tensions, which are derived from a low-pass filtering of the saccadic neural innervation signals, N_j^i. Consistent with Leigh and Zee [1999], each motoneuron innervates a single muscle fiber. The neural

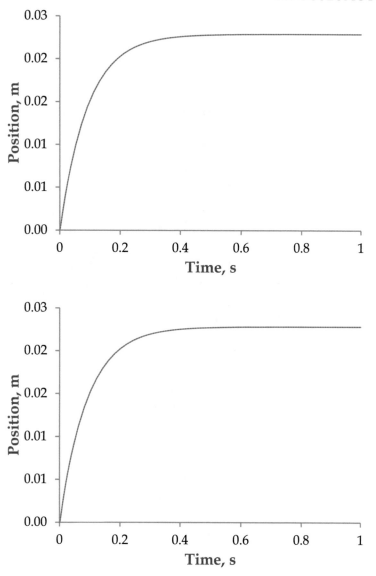

Figure 1.5: Simulations of the whole muscle model and the muscle fiber model contracting with a mass of 0.001 kg (resting on a table), active-state tension 1.4 N and parameters given in Enderle and Zhou [2010]. (Top) Two columns of ten muscle fibers in series (solid line) and the whole muscle model (dots) are illustrated. (Bottom) One thousand muscle fibers in series (solid line) and the whole muscle model (dots) are illustrated. The dotted line is difficult to see since the results are identical. Overall, the simulations for whole muscle are identical to any series or parallel combination of muscle fibers, for any mass or active-state tension.

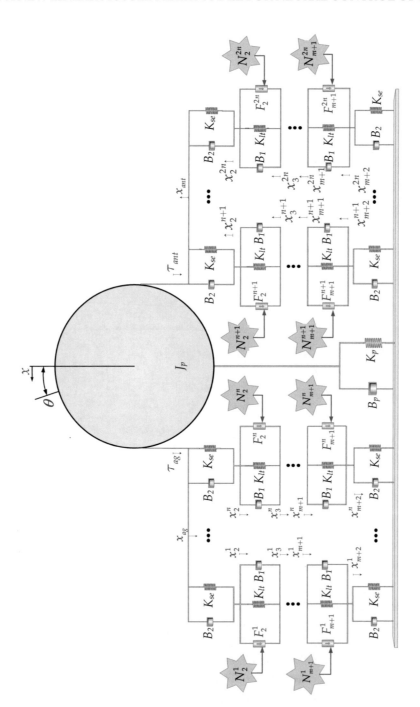

Figure 1.6: Oculomotor plant for horizontal saccades with a muscle fiber muscle model for the agonist and antagonist rectus eye muscles. While not shown, the muscles at primary position are stretched 3.705 mm.

innervation signals are typically characterized as a pulse-step signal, or a pulse-slide-step signal during a saccadic eye movement [Enderle and Zhou, 2010, Goldstein, 1983, Optican and Miles, 1985, Zhou et al., 2009]. In general, the muscle fibers in Fig. 1.6 each can have a separate neural input. The net torques, τ_{ag} and τ_{ant}, generated by the muscles during a saccade rotates the eyeball to a new orientation, and after the saccade is completed, compensates the passive restraining torques generated by orbital tissues.

The tension developed by the agonist muscle is given by

$$\tau_{ag} = \sum_{i=1}^{n} \left(K_{se} \left(x_2^i - x_{ag} \right) + B_2 \left(\dot{x}_2^i - \dot{x}_{ag} \right) \right) \tag{1.20}$$

and by the antagonist muscle is given by

$$\tau_{ant} = \sum_{i=n+1}^{2n} \left(K_{se} \left(x_{ant} - x_2^i \right) + B_2 \left(\dot{x}_{ant} - \dot{x}_2^i \right) \right). \tag{1.21}$$

Using D'Alembert's principle for the agonist muscle that is undergoing shortening, the equations that define the rest of the system are for $i = 1, \ldots, n$

$$
\begin{aligned}
T_i &= -K_{se} \left(x_{ag} - x_2^i \right) - B_2 \left(\dot{x}_{ag} - \dot{x}_2^i \right), \\
K_{se} &\left(x_{ag} - x_2^i \right) + B_2 \left(\dot{x}_{ag} - \dot{x}_2{}^i \right) \\
&= K_{lt} \left(x_2^i - x_3^i \right) + B_1 \left(\dot{x}_2^i - \dot{x}_3^i \right) - F_2^i, \\
K_{lt} &\left(x_2^i - x_3^i \right) + B_1 \left(\dot{x}_2^i - \dot{x}_3^i \right) - F_2^i \\
&= K_{lt} \left(x_3^i - x_4^i \right) + B_1 \left(\dot{x}_3^i - \dot{x}_4^i \right) - F_3^i, \\
&\qquad\qquad \vdots \\
K_{lt} &\left(x_m^i - x_{m+1}^i \right) + B_1 \left(\dot{x}_m^i - \dot{x}_{m+1}^i \right) - F_m^i \\
&= K_{lt} \left(x_{m+1}^i - x_{m+2}^i \right) + B_1 \left(\dot{x}_{m+1}^i - \dot{x}_{m+2}^i \right) - F_{m+1}^i, \\
K_{lt} &\left(x_{m+1}^i - x_{m+2}^i \right) + B_1 \left(\dot{x}_{m+2}^i - \dot{x}_{m+2}^i \right) - F_{m+1}^i \\
&= K_{se} x_{m+2}^i + B_2 \dot{x}_{m+2}^i.
\end{aligned}
\tag{1.22}
$$

Using D'Alembert's principle for the antagonist muscle that is undergoing lengthening, the equations that define the rest of the system are for $i = n + 1, \ldots, 2n$

$$
\begin{aligned}
T_i &= K_{se}\left(x_{ant} - x_2^i\right) + B_2\left(\dot{x}_{ant} - \dot{x}_2^i\right), \\
K_{se}&\left(x_{ant} - x_2^i\right) + B_2\left(\dot{x}_{ant} - \dot{x}_2^i\right) \\
&= K_{lt}\left(x_2^i - x_3^i\right) + B_1\left(\dot{x}_2^i - \dot{x}_3^i\right) + F_2^i, \\
K_{lt}&\left(x_2^i - x_3^i\right) + B_1\left(\dot{x}_2^i - \dot{x}_3^i\right) + F_2^i \\
&= K_{lt}\left(x_3^i - x_4^i\right) + B_1\left(\dot{x}_3^i - \dot{x}_4^i\right) + F_3^i, \\
&\qquad\qquad\vdots \\
K_{lt}&\left(x_m^i - x_{m+1}^i\right) + B_1\left(\dot{x}_m^i - \dot{x}_{m+1}^i\right) + F_m^i \\
&= K_{lt}\left(x_{m+1}^i - x_{m+2}^i\right) + B_1\left(\dot{x}_{m+1}^i - \dot{x}_{m+2}^i\right) + F_{m+1}^i, \\
K_{lt}&\left(x_{m+1}^i - x_{m+2}^i\right) + B_1\left(\dot{x}_{m+1}^i - \dot{x}_{m+2}^i\right) + F_{m+1}^i \\
&= K_{se}x_{m+2}^i + B_2\dot{x}_{m+2}^i.
\end{aligned}
\tag{1.23}
$$

The torques acting on the eyeball are given by

$$
r\left(\tau_{ag} - \tau_{ant}\right) = J_p\ddot{\theta} + B_p\dot{\theta} + K_p\theta.
\tag{1.24}
$$

With $x = \theta r$, or, $\theta = \dfrac{x}{r} \times \dfrac{180}{\pi} = 57.2958\dfrac{x}{r}$, where x is measured in meters with $r = 0.01108$ m, and $J = 57.2958\dfrac{J_p}{r^2}$, $B = 57.2958\dfrac{B_p}{r^2}$, and $K = 57.2958\dfrac{K_p}{r^2}$, Eq. (1.24) is rewritten as

$$
\tau_{ag} - \tau_{ant} = J\ddot{x} + B\dot{x} + Kx.
\tag{1.25}
$$

It should be noted that the muscles are stretched x_p mm from equilibrium at primary position, and that $x = x_{ag} + x_p = x_{ant} - x_p$, $\dot{x} = \dot{x}_{ag} = \dot{x}_{ant}$, and $\ddot{x} = \ddot{x}_{ag} = \ddot{x}_{ant}$.

Rather than solving for the response as in Enderle and Zhou [2010], state variables are used to simplify the system for simulation in MatLab's Simulink. For $i = 1, \ldots, n$, $y_1^i = x_{ag} - x_2^i$ and $y_{m+2}^i = x_{m+2}^i$, and for $j = 2, \ldots, m+1$, $y_j^i = x_j^i - x_{j+1}^i$. It then follows that

$$
x_{ag} = \sum_{j=1}^{m+2} y_j^1 = \ldots \sum_{j=1}^{m+2} y_j^n.
\tag{1.26}
$$

For $i = n + 1, \ldots, 2n : y_1^i = x_{ant} - x_2^i$ and $y_{m+2}^i = x_{m+2}^i$, and for $j = 2, \ldots, m+1$, $y_j^i = x_j^i - x_{j+1}^i$. It then follows that

$$
x_{ant} = \sum_{j=1}^{m+2} y_j^{n+1} = \ldots \sum_{j=1}^{m+2} y_j^{2n}.
\tag{1.27}
$$

For the agonist muscle, note that $y_1^i = x_{ag} - y_2^i - \ldots - y_{m+2}^i$ for $i = 1, \ldots, n$, and with

$$
\begin{aligned}
T_i &= -K_{se} \left(x_{ag} - x_2^i \right) - B_2 \left(\dot{x}_{ag} - \dot{x}_2^i \right) \\
&= -K_{se} y_1^i - B_2 \dot{y}_1^i,
\end{aligned}
\tag{1.28}
$$

the model is given by the following for $i = 1, \ldots, n$,

$$
\dot{y}_1^i = \frac{-\left(T_i + K_{se} y_1^i \right)}{B_2},
\tag{1.29}
$$

and for $j = 2, \ldots, m+1$,

$$
\dot{y}_j^i = \frac{-T_i - K_{lt} y_j^i + F_j^i}{B_1},
\tag{1.30}
$$

and

$$
\dot{y}_{m+2}^i = \frac{-\left(T_i + K_{se} y_{m+2}^i \right)}{B_2}.
\tag{1.31}
$$

The tension generated by the agonist muscle is given by

$$
\tau_{ag} = -\sum_{i=1}^{n} \left(K_{se} y_1^i + B_2 \dot{y}_1^i \right).
\tag{1.32}
$$

For the antagonist muscle, note that $y_1^i = x_{ant} - y_2^i \ldots - y_{m+2}^i$ for $i = n+1, \ldots, 2n$, and with

$$
\begin{aligned}
T_i &= K_{se} \left(x_{ant} - x_2^i \right) + B_2 \left(\dot{x}_{ant} - \dot{x}_2^i \right) \\
&= K_{se} y_1^i + B_2 \dot{y}_1^i,
\end{aligned}
\tag{1.33}
$$

the model is given by the following for $i = n+1, \ldots, 2n$,

$$
\dot{y}_1^i = \frac{T_i - K_{se} y_1^i}{B_2},
\tag{1.34}
$$

and for $j = 2, \ldots, m+1$,

$$
\dot{y}_j^i = \frac{T_i - K_{lt} y_j^i - F_j^i}{B_1},
\tag{1.35}
$$

and

$$
\dot{y}_{m+2}^i = \frac{T_i - K_{se} y_{m+2}^i}{B_2}.
\tag{1.36}
$$

The tension generated by the antagonist muscle is given by

$$
\tau_{ant} = \sum_{i=n+1}^{2n} \left(K_{se} y_1^i + B_2 \dot{y}_1^i \right),
\tag{1.37}
$$

and for the torques acting on the eyeball

$$
\tau_{ag} - \tau_{ant} = J \ddot{x} + B \dot{x} + K x.
\tag{1.38}
$$

1.4   NEURAL INPUT

The neural input to the saccade system is a pulse-slide-step as described in Enderle and Zhou [2010] and Zhou et al. [2009]. This input is consistent with the data published in the literature (for example, see Fig. 4 in Robinson [1981] and Fig. 2 in Van Gisbergen et al. [1981]). The diagram in Fig. 1.7 (Top) closely approximates the data shown in (Bottom) for the agonist input. The antagonist input is illustrated in the (Middle).

At steady-state before the saccade, the eye is held steady by the agonist and antagonist inputs, F_{g0} and F_{t0}. We typically define the time when the target moves as $t = 0$. This is a common assumption since many simulation studies ignore the latent period and focus on the actual movement (see the time axis in Fig. 1.7, Top and Bottom).

The overall agonist pulse occurs in the interval $0 - T_2$, where the start of the pulse occurs with an exponential rise from the initial firing rate, F_{g0}, to peak magnitude, F_{p1}, with a time constant τ_{gn1}. At T_1, the input decays to F_{p2}, with a time constant τ_{gn2}. The slide occurs at T_2, with a time constant τ_{gn3}, to F_{gs}, the force necessary to hold the eye at its destination. The input F_{gs} is applied during the step portion of the input.

At $t = 0$, the antagonist neural input is completely inhibited and exponentially decays to zero from F_{t0} with time constant τ_{tn1}. At time T_3, the antagonist input exponentially increases with time constant τ_{tn2}. The antagonist neural input shown in Fig. 1.7 (Middle) includes a PIRB pulse with duration of $T_4 - T_3$. At T_4, the antagonist input exponentially decays to F_{ts}, with a time constant τ_{tn3}. If no PIRB occurs in the antagonist input, the input exponentially rises to F_{ts} with time constant τ_{tn2}.

The agonist pulse includes an interval (T_1) that is constant for saccades of all sizes as supported by physiological evidence [Enderle, 2002, Enderle and Zhou, 2010, Zhou et al., 2009]. We choose to model the change in the firing rate with an exponential function as this seems to match the data fairly well.

After complete inhibition, the antagonist neural input has a brief excitatory pulse starting at T_3 with duration of approximately 10 ms. Enderle [2002] proposed that this burst is generated by PIRB, a property which contributes to the post-saccade phenomena such as dynamic and glissadic overshoot.

The agonist and antagonist active-state tensions are defined as low-pass filtered neural inputs. Based on the diagram in Fig. 1.7 and assuming the exponential terms reach steady-state at $t = 5\tau$, the equations for each of the agonist neural inputs, N_j^i, for $j = 1, \ldots, m$ and $i = 1, \ldots, n$ is written as

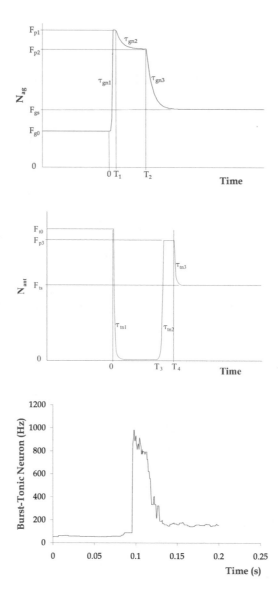

Figure 1.7: Neural input to the horizontal saccade system. (Top) Agonist input. (Middle) Antagonist input. (Bottom) Discharge rate of a single burst-tonic neuron during a saccade (agonist input). Details of the experiment and training for (Bottom) are reported elsewhere [Sparks, D.L., Holland, R., Guthrie, B.L., 1976. "Size and distribution of movement fields in the monkey superior colliculus," *Brain Res.*, 113: 21–34. DOI: 10.1016/0006-8993(76)90003-2.]. (Data provided personally by Dr. David Sparks.)

$$
\begin{aligned}
N_j^i =\,&F_{g0}u(-t)\\
&+\left(F_{g0}+(F_{p1}-F_{g0})\,e^{\frac{-t}{\tau_{gn1}}-5}\right)\times\left(u(t)-u\left(t-5\tau_{gn1}\right)\right)\\
&+F_{p1}\left(u\left(t-5\tau_{gn1}\right)-u(t-T_1)\right)\\
&+\left(F_{p2}+(F_{p1}-F_{p2})\,e^{\frac{(T_1-t)}{\tau_{gn2}}}\right)\times\left(u\left(t-T_1\right)-u\left(t-T_2\right)\right)\\
&+\left(F_{gs}+\left(\begin{array}{c}F_{p2}+(F_{p1}-F_{p2})e^{\frac{(T_1-T_2)}{\tau_{gn2}}}\\-F_{gs}\end{array}\right)e^{\frac{(T_2-t)}{\tau_{gn3}}}\right)\times u\left(t-T_2\right).
\end{aligned}
\tag{1.39}
$$

The agonist active-state tension for $j = 1,\ldots, m$ and $i = 1,\ldots, n$ is written as

$$
\dot{F}_j^i = \frac{N_j^i - F_j^i}{\tau_{ag}}.
\tag{1.40}
$$

The antagonist neural inputs, N_j^i, for $j = 1,\ldots, m$ and $i = n+1,\ldots, 2n$ is written as

$$
\begin{aligned}
N_j^i =\,&F_{t0}u(-t)\\
&+F_{t0}e^{\frac{-t}{\tau_{tm1}}}\left(u(t)-u(t-T_3)\right)\\
&+\left(F_{t0}e^{\frac{-T_3}{\tau_{tm1}}}+\left(F_{p3}-F_{t0}e^{\frac{-T_3}{\tau_{tm1}}}\right)e^{\frac{-(t-T_3)}{\tau_{tm2}}-5}\right)\\
&\quad\times\left(u\left(t-T_3\right)-u\left(t-T_3-5\tau_{tm2}\right)\right)\\
&+F_{p3}\left(u\left(t-T_3-5\tau_{tm2}\right)-u(t-T_4)\right)\\
&+\left(F_{ts}+(F_{p3}-F_{ts})\,e^{\frac{(T_4-t)}{\tau_{tm3}}}\right)u(t-T_4).
\end{aligned}
\tag{1.41}
$$

The antagonist active-state tension for $j = 1,\ldots, m$ and $i = n+1,\ldots, 2n$ is written as

$$
\dot{F}_j^i = \frac{N_j^i - F_j^i}{\tau_{ant}}.
\tag{1.42}
$$

Note that Eqs. (1.39) and (1.41) are written in terms of intervals. Further, we assume that $5\tau_{gn1} < T_1$ and $T_3 + 5\tau_{tm2} < T_4$, which simplifies analysis, where

$$
\tau_{ag} = \tau_{gac}\left(u(t)-u\left(t-T_2\right)\right)+\tau_{gde}u\left(t-T_2\right),
\tag{1.43}
$$

and

$$
\tau_{ant} = \tau_{tde}\left(u(t)-u\left(t-T_3\right)\right)+\tau_{tac}\left(u\left(t-T_3\right)-u\left(t-T_4\right)\right)+\tau_{tde}u\left(t-T_4\right).
\tag{1.44}
$$

1.5   RESULTS

Parameter estimation is carried out as previously described using the system identification technique [Enderle and Zhou, 2010, Zhou et al., 2009]. The accuracy of this method is excellent. Parameters for the muscle fiber model are calculated using Eq. (1.7) and the estimates previously found, adjusted with $n = 1$ and $m = 100$. Using 100 identical muscle fibers provides sufficient resolution to test our hypothesis, and the same result would occur if we used 10 columns of 10 muscle fibers in series. Increasing the number of muscle fibers increases the accuracy between the whole muscle oculomotor model and the muscle fiber oculomotor model. The estimates for the other parameters remain unchanged. The data analyzed is for five horizontal saccades $(4°, 8°, 12°, 16°,$ and $20°)$, recorded from two medium lead burst neurons, one long lead burst neuron, and one burst tonic neuron. A total of 20 saccades are analyzed.

Initial conditions for the system are determined using Eqs. (1.8)–(1.12), here, computing the state variables and tensions at time zero, rather than at $t = \infty$. The muscles are assumed to be stretched by 3.705 mm at primary position [Enderle and Zhou, 2010, Zhou et al., 2009].

For simplicity in the muscle fiber model, we use identical neural inputs for the agonist muscle using Eq. (1.39), and the antagonist muscle using Eq. (1.41). The agonist and antagonist active-state tensions are computed using Eqs. (1.40) and (1.42), respectively. It should be noted that we have used one column of muscle fibers for each muscle, and that if we used multiple columns of muscle fibers, the neural input to each muscle fiber needs to be scaled by $\frac{1}{n}$, with virtually no change in the results.

To begin the analysis of the muscle fiber oculomotor plant, we use the parameter values previously estimated for horizontal monkey saccades and focus only on the 4° and 8° saccades [Enderle and Zhou, 2010, Zhou et al., 2009]. As noted previously, there is considerable variation in the estimated neural input for the 12°, 16°, and 20° sacades, and this input captures the firing rate of all the active neurons in a single neuron. It is expected that the number of active neurons varies from saccade to saccade and this explains some of the dynamics observed in the main sequence diagram as described later in the chapter.

To estimate the number of active neurons for the 4° and 8° saccades, the average estimate of the pulse magnitudes, F_{p1} and F_{p2}, from all of the 12°, 16°, and 20° saccades are used to define the input for the 4° and 8° saccades for the muscle fiber model. All the other parameters remain unchanged. To estimate the number of active neurons for the 4° and 8° saccades, the number of active neurons is reduced from 100 until the position estimate simulated by the muscle fiber oculomotor model matches that of the whole muscle oculomotor model. For each of the agonist neurons that are not activated (or turned off), the tonic neuron activity exponentially decays and rises using the time constants for the agonist muscle. The antagonist neuron behavior remains unchanged.

The MatLab/Simulink simulation model for the muscle fiber oculomotor model is shown in Fig. 1.8. In Fig. 1.8a, the overall system model is shown based on Eq. (1.38), which is shown as the upper block in Fig. 1.8b illustrating the overall model with the agonist muscle on the lower left and the antagonist muscle on the lower right. Shown in Fig. 1.8c is the agonist muscle with each block representing five muscle fibers, with each of the five muscle fibers shown in Fig. 1.8d based on Eq. (1.8). One hundred muscle fibers are part of the agonist and antagonist muscles. Shown in Fig. 1.8e is the agonist neural input based on Eq. (1.39), with each set of elements representing a line in Eq. (1.39). The antagonist neural input, defined by Eq. (1.41), is shown in Fig. 1.8f, where each set of elements representing a line in Eq. (1.41). In Fig. 1.8g, the agonist active-state tension is shown on the top according to Eqs. (1.40) and (1.43). The agonist tonic activity for those neurons turned off is shown in the lower portion of Fig. 1.8g. The antagonist active-state tension is shown in Fig. 1.8h based on Eqs. (1.42) and (1.44).

Shown in Fig. 1.9 are the 4° and 8° saccade simulation results for one of the medium lead burst neurons, with both the muscle fiber (solid line) and whole muscle (dots) oculomotor model simulation results plotted on the same graph. For the active neurons, each of the muscle fibers is stimulated with the average F_{p1} and F_{p2} for all saccades and estimating the number of active neurons as previously described. As shown, the results shown demonstrate an excellent match between the whole muscle and muscle fiber oculomotor plant. The match between the two models can be improved by increasing the number of muscle fibers since a finer resolution is possible. The excellent accuracy of these results against the original data remains the same as reported previously [Enderle and Zhou, 2010, Zhou et al., 2009].

Figure 1.10 shows the agonist neural input for 4° and 8° saccades from medium lead burst neuron 1. For the 4° saccade, 48 neurons fire with this neural input, and 76 neurons fire with the 8° neural input. Each of the agonist neurons fire maximally during the pulse phase of the input, rising to the same peak firing rate, and then decaying to a plateau (the only difference between the two graphs is the duration of the burst).

Shown in Fig. 1.11 is the percentage of active agonist neurons as a function of saccade magnitude for the 4° and 8° saccades. Quite clearly, the number of active neurons increases from 4° to 8° as expected since the pulse magnitude increases. It should be clear that the larger the estimate of the original agonist pulse magnitude, the greater the number of active neurons. Shown in Fig. 1.12 are the monkey data firing rates for a medium lead burst neuron for saccades of 4°, 8°, 12°, 16°, and 20°, which are consistent with others published in the literature (for example, see Fig. 4 in Robinson [1981] and Fig. 2 in Van Gisbergen et al. [1981]). The maximum neuron firing rate during the early part of the saccade is quite variable, and does not appear to be a function of saccade size. In general, only the duration of the pulse affects the size of the saccade. The simulation results in this study are consistent with the data(all of the active agonist neurons fire maximally during the pulse for the 4° and 8° saccades.

One can extend this analysis to the other saccades of 12°, 16°, and 20° by either using the maximum pulse magnitude for that particular neuron, or for all neurons. Here, we use the maximum F_{p1} and F_{p2} for that particular neuron to estimate the number of active neurons for the 12°, 16°, and 20° saccades, with all other parameters left unchanged as before. Shown in Fig. 1.13

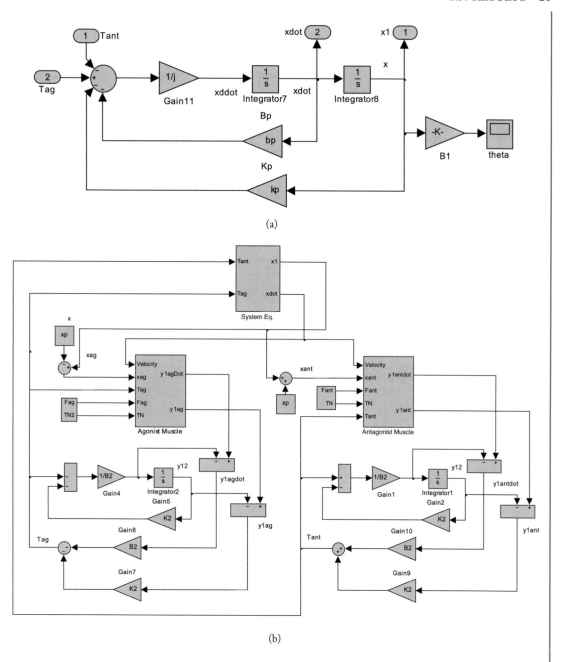

Figure 1.8: (a) System model based on Eq. (1.37). (b) High level model of the agonist and antagonist muscles based on Eqs. (1.28)–(1.35) and system model at the top. *(Continues.)*

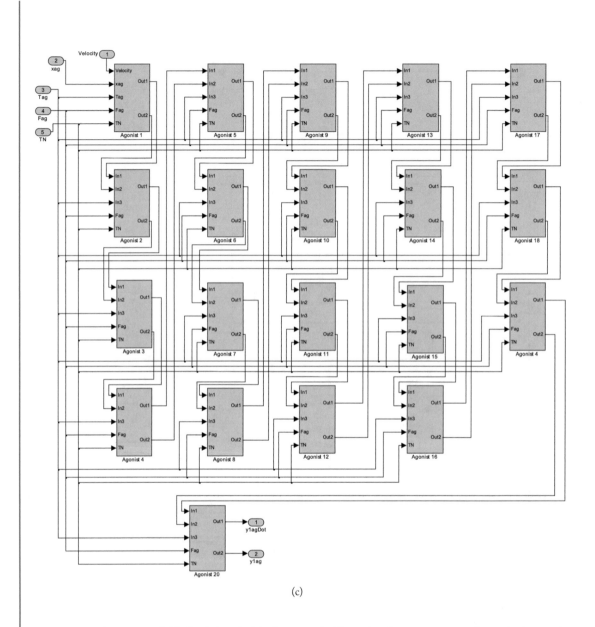

(c)

Figure 1.8: *(Continued.)* (c) Each box holds five muscle fibers for the agonist muscle, with the total of 20 boxes representing 100 muscle fibers. The antagonist is similar in structure. *(Continues.)*

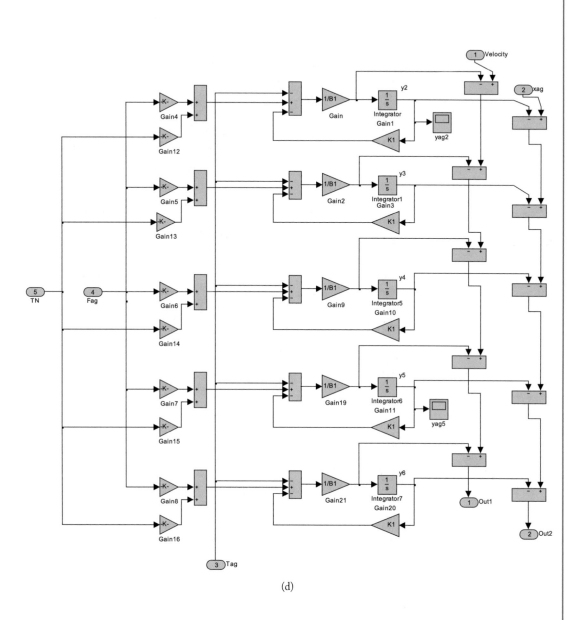

(d)

Figure 1.8: *(Continued.)* (d) The five muscle fibers contained within each box in Fig. 1.8c. *(Continues.)*

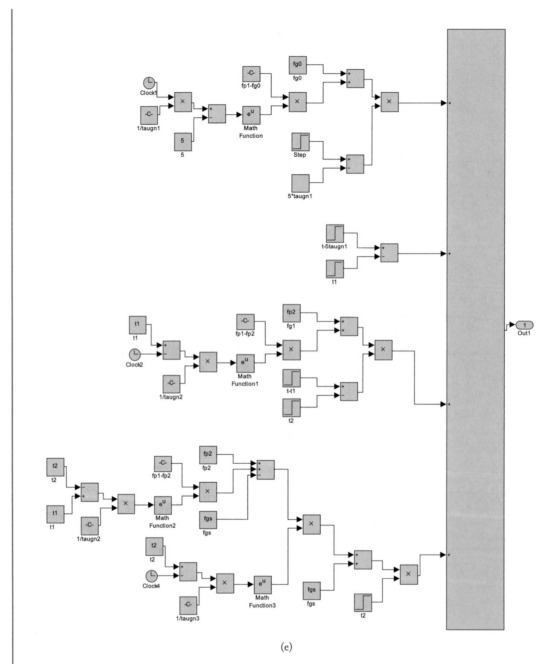

(e)

Figure 1.8:   *(Continued.)* (e) Agonist neural input. *(Continues.)*

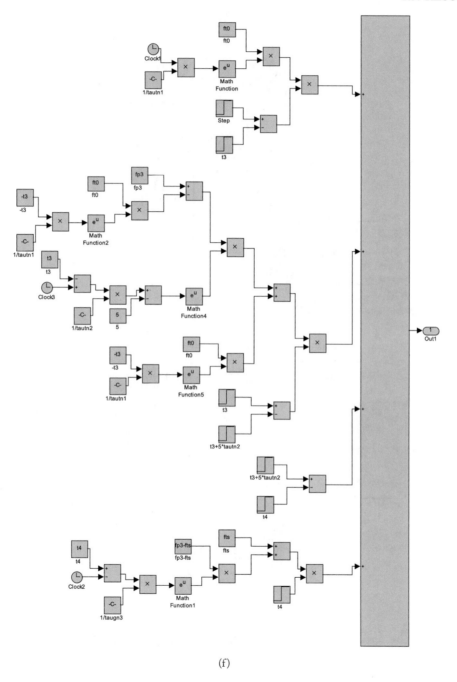

(f)

Figure 1.8: *(Continued.)* (f) Antagonist neural input. *(Continues.)*

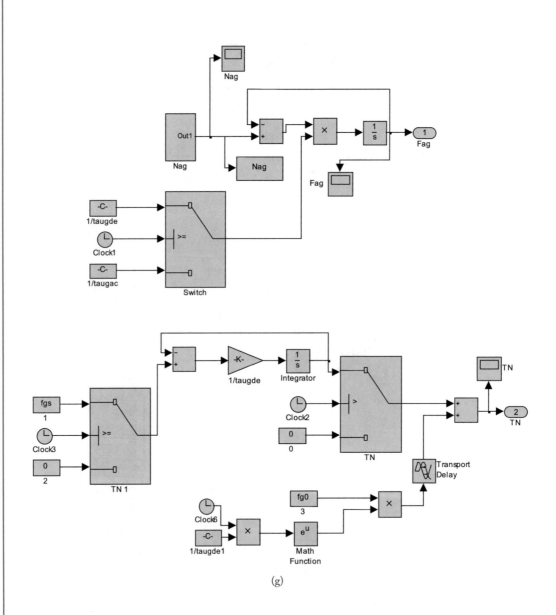

(g)

Figure 1.8: *(Continued.)* (g) Agonist active-state tension (Top). Agonist tonic input (Bottom). *(Continues.)*

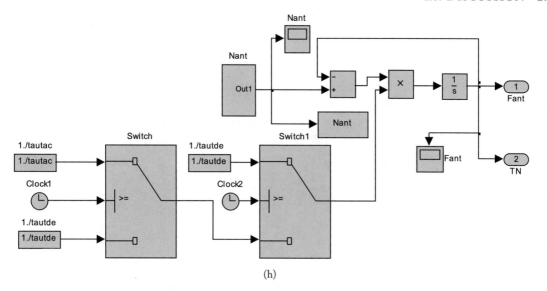

Figure 1.8: *(Continued.)* (h) Antagonist active-state tension.

are the simulation results for a medium lead burst neuron. As shown, the results demonstrate an excellent match between the whole muscle and muscle fiber oculomotor model. The other three neurons exhibit similar results.

1.6   DISCUSSION

In our previous effort, we summed the input of all active motoneurons into the firing of a single neuron in our parameter estimation effort [Enderle and Zhou, 2010, Zhou et al., 2009]. Even though our simulation results accurately matched the data, we knew that the amplitude dependent agonist input for small saccades is non-physiologic for individual neurons, that is, the firing rate of a real agonist neuron is very high and does not change as a function of saccade magnitude as easily seen in Fig. 4 in Robinson [1981], Fig. 2 in Van Gisbergen et al. [1981], and our Fig. 1.11 data. As the magnitude of the saccades increases, the firing rate of the single neuron in our previous agonist controller model increases up to 8°, after which it remains at a maximal level of firing. The overall neural input for the agonist pulse is given by

$$N_{ag} = \begin{cases} N\,(\theta_T)\,N_{ag_i} & \theta < 8° \\ N_{ag_{\max}} & \theta \geq 8° \end{cases}, \tag{1.45}$$

where $N(\theta_T)$ is the number of neurons firing for a saccade of θ degrees, N_{ag_i} is the contribution from an individual neuron, and $N_{ag_{\max}}$ is the combined input from all neurons.

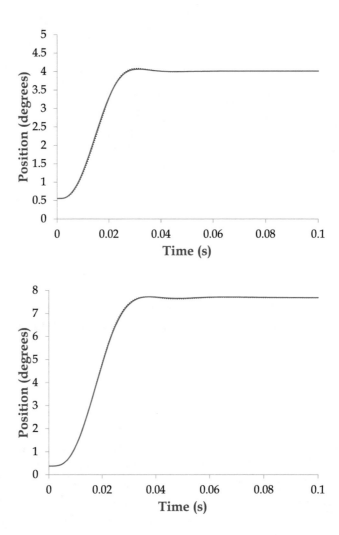

Figure 1.9: Simulation results for the 4° and 8° saccades for the medium lead burst neuron 1, with both the muscle fiber (solid line) and whole muscle (dots) oculomotor plant simulation results plotted on the same graph.

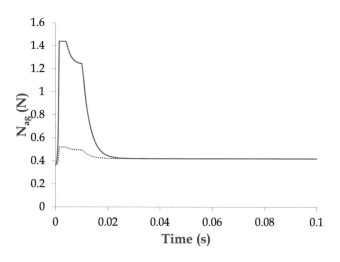

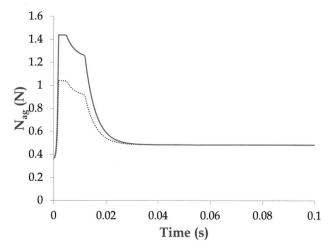

Figure 1.10: Agonist neural input for 4° (Top) and 8° (Bottom) saccades for a medium lead burst neuron. The neural input from the muscle fiber model is shown in the solid line and the whole muscle is shown in the dotted line.

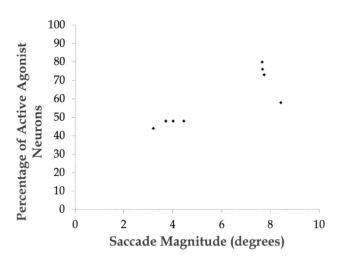

Figure 1.11: Percentage of active agonist neurons as a function of saccade magnitude for 4° and 8° saccades for the four neurons.

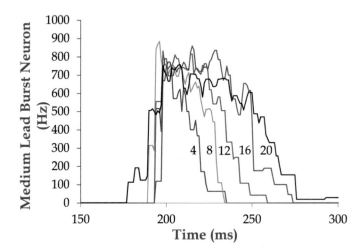

Figure 1.12: The firing rate for one of the medium lead burst neuron 1 for saccades of 4°, 8°, 12°, 16°, and 20°.

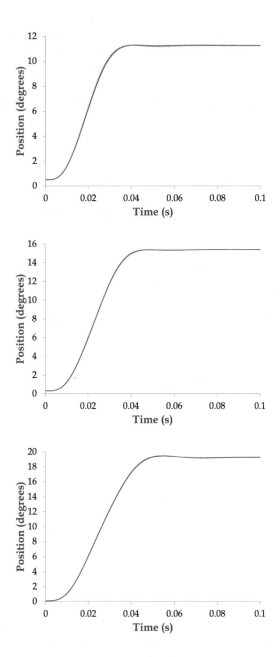

Figure 1.13: Simulation results for the 12°, 16°, and 20° saccades for a medium lead burst neuron. The muscle fiber (solid line) and whole muscle (dots) oculomotor model simulation results are plotted on the same graph. The percentage of neurons actively firing is: 75% for the 12°, 100% for the 16°, and 92% for the 20° saccade.

The objective of this chapter is to investigate the importance of the number of active neurons driving the eyes during a saccade, and its implications in the control of saccades. To accomplish this, a new linear muscle fiber model is presented that has the same characteristics of the whole muscle model described in Enderle and Zhou [2010] and Zhou et al. [2009], that is, it exhibits accurate nonlinear force-velocity and length-tension relationships. Moreover, a one-to-one parametric relationship between the two models exists that is scalable—a 1,000 muscle fiber model, involving multiple configurations of muscle fibers in series or parallel, performs identically to a whole muscle model.

A muscle fiber-based oculomotor plant driven by a realistic neural input is presented that performs accurately for saccades of all sizes using a time-optimal controller. This chapter quantifies the time-optimal controller described previously by Enderle and coworkers [Enderle, 2002, Enderle and Zhou, 2010, Zhou et al., 2009], where agonist neurons fire maximally during the agonist pulse and independent of eye orientation, while the antagonist muscle is inhibited. As shown, only the duration of the agonist pulse and the number of agonist neurons firing maximally determines the saccade size. These observations more fully supports a time-optimal controller for the saccade system, with a single switch time using a realistic pulse-slide-step motoneuron stimulation of the agonist muscle with a pause and step in the motoneuron stimulation of the antagonist muscle, and physiological constraints. The time-optimal controller still operates in two modes, one for small saccades and one for large saccades.

The parameter estimates of this study continue to support the main sequence diagram in agreement with other studies. The peak velocity has an exponential shape as a function of saccade magnitude, the latent period is rather independent of saccade magnitude, and saccade duration is linearly related to saccade magnitude for saccades over 8°, and independent of saccade magnitude for saccades under 8° for the monkey data examined here.

1.6.1   MAXIMAL FIRING OF THE AGONIST NEURONS

For most investigators, a disconnect exists between the firing rate in agonist neurons and saccade magnitude. For example, a velocity-based controller, such as the one proposed by Sylvestre and Cullen [1999], is not supported by our analysis of agonist burst firing observed in the data. For instance, all of the neurons in Fig. 1.11 seem to arrive at approximately the same high peak firing rate, and then descend to a plateau firing rate, and finally descend to a step firing rate necessary to keep the eyes at their destination. The most notable difference in the firing rate is the duration of the pulse.

Moreover, experimental data do not show a one-to-one relationship between EBN firing rate and saccade magnitude in our view. Other saccade generator models, such as the Scudder model [1988] and the Gancarz and Grossberg model [1998], are structured to provide a firing rate-saccade amplitude dependent signal. Cullen and coworkers [1996] used a system identification technique and found a firing-rate, saccade-amplitude-dependent controller. None of these studies used a homeomorphic oculomotor saccade plant or considered a population of neurons.

Our results continue to demonstrate that a first-order time-optimal controller is sufficient to generate saccades that are tightly coupled with appropriate main sequence characteristics.

1.6.2   SACCADE AND AGONIST PULSE DURATION

In the exercises carried out in this project, all time-based parameter estimates are kept the same as previously estimated. Only the pulse magnitude is changed to a maximum rate for saccades of all sizes when the number of actively firing agonist neurons is included in the model. Thus, all comments related to the pulse duration and saccade duration hold true from our previous work [Enderle and Zhou, 2010, Zhou et al., 2009]. That is, small saccades all have approximately the same duration and the duration does not change as a function of saccade magnitude(it is the number of neurons that actively fire that determines the size of the saccade. For large saccades, saccade duration is directly related to saccade magnitude.

The same finding applies to the agonist pulse duration. Our hypothesis concerning small saccade pulse duration is that there is a minimum time period that the excitatory burst neuron (EBN) can be switched on and off (approximately 10 ms), and that this is a physical constraint of the system (biophysical property of the neuron membrane). This is also supported by the work of Hu and coworkers [2007], who examined the variability in saccade amplitude, duration, and velocity, and the reliability of the EBN firing rate in the monkey. For saccades with similar amplitude and velocity, it was determined that the initial portion of the EBN firing rate had little variability, while the last portion of the burst had observable variability.

1.6.3   NUMBER OF ACTIVE NEURONS AND TIME-OPTIMAL CONTROL

Given that all active neurons fire maximally during the agonist pulse, it is the duration of the agonist pulse and the number of actively firing agonist neurons that determines the size of the saccade according to our time-optimal controller. While there is variability in the estimates of the pulse magnitude, the 12° saccades generally have fewer active neurons than the 16° and 20° saccades, but the neurons are all firing maximally. The 16° and 20° saccades have the same general pulse magnitude characteristics.

As described, the superior colliculus is arranged in a movement field based on the saccade magnitude [Enderle and Zhou, 2010]. Sparks and coworkers [1976] described increasing movement field of activity in the superior colliculus up to 10°, which then remained constant for those greater than 10°. Based on our findings here for the monkey data, it may be possible that the movement field continues to increase up to 12° rather than 10°, as previously reported. This may also be a function of the maximal burst firing rate of the EBN when released from inhibition and the ability to turn off the EBN firing during the slide portion of the agonist controller. This characteristic was not evident in the human pulse magnitude estimates from previous analysis, the estimated pulse magnitudes are not a function of saccade magnitudes for saccades greater than 7°.

One stark feature is that monkey has a larger steady-state peak velocity than human in the main sequence peak velocity-saccade magnitude relationship, reported at 694.64°/s with time constant 8.7 s for monkey and 400°/s with time constant of 5 s for human. This difference may be due to the an increasing number of active agonist neurons up to 12°.

These results are still consistent with the time-optimal controller previously described, where the number of active agonist neurons participating in the saccade continue to increase up to 12°, and then remained constant for those saccades greater than 12° in monkey. While this is a minor change from our previous hypothesis, the controller is still quite simple—the cerebellum keeps track of the duration of the agonist pulse and the number of agonist neurons actively firing in coordinating the end of the saccade, and not the firing rate of individual neurons. Continued focus on human is certainly warranted.

1.6.4   SYNCHRONY OF AGONIST NEURON FIRING

While there is an infinite number of combinations for the start time for each of the of the active agonist neurons, we chose to assume that all active neurons fired identically and varied the start time to investigate synchrony of firing and kept all the other parameter estimates the same. Smaller saccades have fewer actively firing agonist neurons than larger saccades. Thus, if the same number of neurons is delayed, then the affect should be more noticeable for the smaller saccade than the larger one. Shown in Fig. 1.14 are two simulations of a 4° and 20° saccade, each with 10 active neurons delayed by 5 ms. It is clear that the 4° saccade is more affected than the 20° saccade, since approximately 20% of the 4° neurons are delayed as opposed to 10% of the 20° neurons. This is especially noticeable in the later portion of the saccade. For the 4° saccade, the duration is longer and peak velocity is lower for the delayed than the non-delayed 4° saccade. This impact is much less for the 20° saccade. The impact for the 8°, 12°, and 16° are similarly affected based on the number of active agonist neurons.

1.6.5   VARIABILITY IN AGONIST NEURON FIRING

The variability in the firing rate in an individual neuron and the effect on saccade accuracy is also an objective of this study. Sparks and coworkers [1976] describe the variability of neurons firing as follows:

> "The precision or accuracy of a saccade results from the summation of the movement tendencies produced by the population of neurons rather than the discharge of a small number of finely tuned neurons. The contribution of each neuron to the direction and amplitude of the movement is relatively small. Consequently, the effects of variability or 'noise' in the discharge frequency of a particular neuron are reduced by averaging over many neurons. By reducing the effects of 'noise' in the discharge frequency of individual neurons, the large movement fields (which result in large populations of neurons being active during a specific movement) may contribute to, rather than detract from, the accuracy of the saccade."

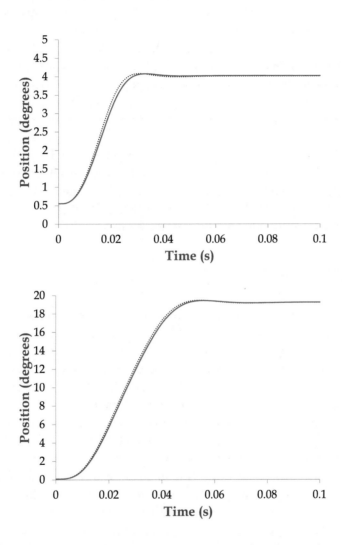

Figure 1.14: Simulation results for the 4° and 20° saccades with a delay of 0.005 ms for 10 muscle fibers for a medium lead burst neuron. The muscle fiber (solid line) and whole muscle (dots) oculomotor model simulation results are plotted on the same graph. The number of neurons actively firing is: 48 for the 4° and 92 for the 20°.

Clearly, the variability in the individual firing rates of the active agonist neurons will impact the size of the saccade and its dynamics, but this affect is rather small. To illustrate this, consider the Top graph in Fig. 1.15, which shows the simulation results generated with the muscle fiber model when 20 agonist neurons are firing 15% higher and 20 agonist neurons are firing 15% lower than the rate reported previously for a 20° saccade for the medium lead burst neuron (solid line) in Fig. 1.13 against the whole muscle oculomotor plant firing normally (dots). There is no affect whatsoever observed in this situation—the average plus magnitude over the entire population of actively firing agonist neurons essentially drives the eyes to their destination in agreement with the postulate by Sparks and coworkers. Here the average pulse magnitude equals the original pulse magnitude.

Next consider the case in which the average pulse magnitude differs from the original pulse magnitude for the whole muscle oculomotor model, as illustrated in the bottom graph in Fig. 1.15. Here, there are 20 agonist neurons firing 15% lower than the rest of the neurons, which results in the average pulse magnitude lower than the original pulse magnitude. The amplitude of the saccade is less, as is the peak velocity. The duration is unaffected.

1.7   CONCLUSION

A muscle fiber muscle model incorporated into oculomotor plant under time-optimal control is presented to describe the horizontal saccadic eye movement system. The muscle fiber muscle model provides an excellent match to the data and has the same characteristics of our previous whole muscle model in that it reproduces the nonlinear force-velocity and length-tension relationships observed in data [Enderle et al., 1991]. A tendon at each end of the muscle connected between columns of muscle fibers connected in series form the muscle. The tendon consists of a viscous and elastic element in parallel, and the muscle fiber consists of an active-state generator in parallel with a viscous and elastic element. The muscle fiber muscle model is scalable with the whole muscle model and performs identically with the whole muscle model whether the muscle fibers are connected in series or parallel. The importance of the muscle fiber muscle incorporated into the oculomotor plant is that it allows us to investigate the importance of the firing of individual neurons and the number of actively firing neurons in the control of saccades.

The muscle fiber oculomotor model is parameterized using previous estimates of the parameters, appropriately scaled, from the whole muscle oculomotor saccade model [Enderle and Zhou, 2010, Zhou et al., 2009]. The muscle fiber oculomotor model still provides an excellent fit between the data and the model predictions as previously described.

For demonstration purposes, the agonist and antagonist muscles each consisted of 100 muscle fibers in series in the new oculomotor model, which provides enough resolution for our study. Each muscle fiber is connected to an individual neuron. For simplicity, the agonist and antagonist neurons all fired identically when investigating the time-optimal control of saccades. For small saccades, the magnitude of the agonist pulse for each neuron is the average value of all

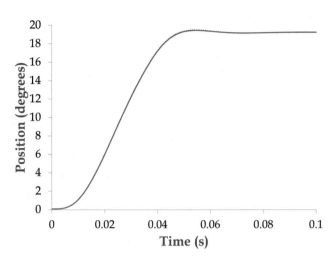

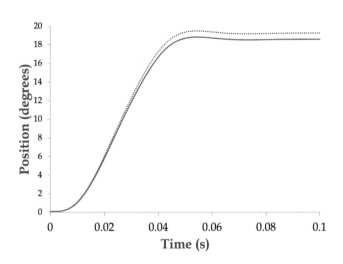

Figure 1.15: Simulation results for the 20° saccade from a medium lead burst neuron. (Top) Simulation results for 20 neurons firing at a rate 15% higher and 20 neurons firing at a rate 15% lower for the muscle fiber oculomotor model (solid line), plotted against the whole muscle oculomotor model (dots) firing normally. (Bottom) Simulation results for 20 neurons firing at a rate 15% lower for the muscle fiber oculomotor model (solid line), plotted against the whole muscle oculomotor model (dots) firing normally.

the larger saccades—as demonstrated, the number of neurons firing maximally is a function of saccade size in agreement with the time-optimal controller.

Simulations with the muscle fiber oculomotor model under a variety of conditions are explored that demonstrate an excellent match between the saccade position data and the model predictions for monkey data. Additionally, the estimates of the neural controller are in agreement with the neuron data. Synchrony of agonist neuron firing has a larger effect for small saccades than larger ones, and variability in agonist neural firing rate has a minor effect on saccades. These are important filters in the control of saccades.

The literature supports a maximal firing frequency of neurons during small saccades, which was not directly supported in the previous study. With the muscle fiber model, the number of active neurons firing maximally is an important parameter that allows the neural controller to match the data for small saccades.

CHAPTER 2

A Physiological Neural Controller of a Muscle Fiber Oculomotor Plant in Horizontal Monkey Saccades[1]

2.1 INTRODUCTION

The control mechanism of the human binocular vision is staggering in its complexity, and has stunned many neuroscientists in their quest to match its adaptive functionality to a greater or lesser degree. As one of the increasingly explored conjugate eye movements, saccades are known as rapid shifts in the gaze direction, wherein a target is tracked by registering its image and peripheral on the fovea. After the detection of this complex of the targets and its peripheral by the visual system, the target is recognized from the peripheral, its position is determined, and consequently, whether or not the corresponding eye movement is a saccade is decided. The fact that it takes 15–100 msec for the visual system to identify a saccade in such a paradigm (timing) has continued to be an astonishing drive in the eye movement research. Indeed, the manner in which synchrony in activity of voltage- and time-dependent cell conductances in a population of neurons triggers motoneurons to lead to a saccade remains fascinating. Each saccade is provoked when a sequence of coordinated activities arise in the midbrain, by the end of which motoneurons drive the lateral and medial rectus muscles.

In the interest of a better understanding of the steps that are significant to the saccade initiation and flow in mesencephalic neural pathways, there have been advances in the involved computational neural modeling. These advances have supplied us with abundant information at different structural scales, such as the biophysical, the circuit and the systems levels. A synergistic use of the experimental (tied with neural data) and computational approaches has provided more coherence and uniformity across the exploration of the midbrain circuitry responsible for saccade generation. The emerging insight from such synergistic approaches, along with the relative convenience at which one can characterize the interaction with the oculomotor periphery,

---

[1]Some of the material in this chapter is an expansion of a previously published paper: Ghahari, A., and Enderle, J.D. (2014) "A Physiological Neural Controller of a Muscle Fiber Oculomotor Plant in Horizontal Monkey Saccades," *ISRN Ophthalmology*, Article ID 406210. DOI: 10.1155/2014/406210.

has steered researches to dealing with the midbrain core drive impact to the saccade generation. Furthermore, new computational predictions could give rise to unexplored experimental tests to study sensorimotor control. The research continues to propose more agreeable findings from the low level of the cellular mechanism of synaptic transmission, to the high level evidence of the connection between the complexes of interconnected neurons.

A physiologically valid system of eye movement relies on the definition of a neural network, a neural controller, and an oculomotor system, all of which has to conform to physiological constraints. The saccade neural network requires involvement of a series of neurons designed to mimic the behavior of actual neuronal populations in the horizontal saccade controller. A generic neuron model is therefore desired to approximate the saccade-related neural activity, thus reflecting the physiology linked to the dendrite, cell body, axon, and presynaptic terminal of each neuron. The continuing research effort in demonstrating such a model has been driven by the need to provide the means to develop a network of neurons, tailored to the complexity involved with inherent physiological evidence. To encompass all of the desired neural behaviors for the other neurons, several modifications to the generic neuron model seem necessary that directly impact its firing rate trajectory [Enderle, 2002, Enderle and Zhou, 2010, Zhou et al., 2009].

The widespread use of spiking neural networks (SNNs) stems from leveraging efficient learning algorithms to the spike response models [Ghosh-Dastidar and Adeli, 2007, 2009]. A spike pattern association neuron identified five classes of spike patterns associated with networks of 200, 400 and 600 synapses, with success rates of 96%, 94%, and 90%, respectively [Mohemmed et al., 2012]. Rosselló et al. presented a hybrid analog-digital circuitry to implement an SNN, reproducing the postsynaptic potential by integrating the filtered action potentials [2009]. A brainstem saccadic circuitry, corroborated by several contributions of local field potentials (LFPs) to the dynamics of neuronal synaptic activity between three neural populations in generating horizontal and vertical saccades in two rhesus monkeys, was introduced by Van Horn et al. [2010]. The extracellular recordings, including spike trains and LFPs, were taken from the saccadic burst neurons (SBNs) in the paramedian pontine reticular formation (PPRF) at the premotor level, the omnipause neurons in the nucleus raphe interpositus, and the motoneurons at the motor level. It was concluded that LFPs from each neuron encode the eye velocity in both the ipsilateral and contralateral directions. In addition, LFP response amplitude of the SBNs was described as a function of saccade direction (in 400 saccades) by fitting Gaussian curves to data (see Fig. 8B in Van Horn et al. [2010]), indicating that the SBN LFPs can be fine-tuned over all the directed saccades. A neural system comprised of a persistent firing sensory neuron, a habituating synapse, and a motoneuron was developed to illustrate the spike-timing dependency of the working memory [Ramanathan et al., 2012]. The persistent firing neuron stems from the Izhikevich neuron model [2003], the habituating synapse is a conductance-based model, and the motor neuron follows the Hodgkin Huxley (HH) model [1952b]. These studies provide abundant evidence that an SNN is well suited to evoke the properties of the firing patterns of the premotor neurons during the pulse and slide phases of a saccade. However, none of the studies have presented a demonstration of the neural circuits reproducing electrophysiological responses in a network of neurons at both premotor and motor levels. To encompass all of the desired neural behaviors, a neural circuitry is used to match the firing rate trajectory of the premotor neurons [Enderle and Zhou, 2010]. We model the saccade-induced spiking activities at the premotor level with an HH model for the bursting neurons and with a modified FitzHugh-Nagumo (FHN) model for the tonic spiking neurons [Faghih et al., 2012].

Time-optimal control theory of the horizontal saccades establishes the fact that there is a minimum time required for the eyes to reach their destination by involving thousands of neurons. Conjugate goal-directed horizontal saccades were well characterized by a first-order time-optimal neural controller [Enderle and Zhou, 2010]. It is important that this new, more complex time-optimal controller ascertains that the firing rate of the motoneurons does not change as a function of saccade magnitude during the pulse innervation of the oculomotor plant.

The kinematics of ocular rotation is relatively simple when compared to that of limb movements, so the oculomotor plant is suited to study the sensorimotor control at a less tedious level. As elaborated in the previous chapter, the muscle fiber model (MFM) improves the oculomotor plant by using several configurations of muscle fibers in series or parallel to drive the eyes to their destination. In other words, it elevates the whole muscle model by Enderle et al. [1991] to the level of muscle fiber model by calculating the viscosities and elasticities of the latter model in terms of the parameter values in the former model. As demonstrated in Section 1.6, increasing the number of muscle fibers results in a closer saccadic agreement between the two muscle models. It is indicated that the muscle fiber model substantiates the fact that the number of motoneurons firing has the highest influence in the accuracy of saccade controller, contradictory to the control strategy of adjusting the firing rates among the whole neurons. Investigation of muscle fiber model is advantageous because it allows for recognizing the effects of the firing of individual neurons, as well as the number of active neurons firing maximally, in controlling the saccades. This investigation also provides an optimum fit for the agonist and antagonist neural controllers to match the experimental data for the small saccades.

In this chapter, we focus on neural control of horizontal monkey saccades. A neural network model of saccade-related neural sites in the midbrain is presented first. We next characterize the underlying dynamics of each neural site in the network, which needs to be treated in the case of spiking neurons. Consequently, to match the dynamics of the neurons and their synaptic network function, a saccadic circuitry, including omnipause neuron (OPN), premotor excitatory burst neuron (EBN), inhibitory burst neuron (IBN), long lead burst neuron (LLBN), tonic neuron (TN), interneuron (IN), and motoneurons of abducens nucleus (AN), and oculomotor nucleus (ON), is developed. The computational neural modeling is motivated by discussing the general applicability of SNNs to the biophysical modeling of interconnected neurons. This perspective elucidates broad insights to modeling at different structural scales, such as the circuit and the systems levels, which we examine subsequently. Finally, the motoneuronal control signals drive a time-optimal controller that stimulates the muscle fiber oculomotor plant.

For a demonstration of the performance responses of the human and monkey saccadic systems, we introduce a graphical user interface (GUI) in the end. The "conjugate goal-directed horizontal monkey saccade" is abbreviated with the term "saccade" throughout this chapter. The terms "motoneurons" and "agonist (antagonist) neurons" are also substitutable.

2.2 NEURAL NETWORK

Neurophysiological evidence and developmental studies indicate that important neural populations, consisting of the cerebellum, superior colliculus (SC), thalamus, cortex, and other nuclei in the brainstem, are involved in the initiation and control of saccades [Coubard, 2013, Enderle, 1994, Enderle and Engelken, 1995, Enderle, 2002, Enderle and Zhou, 2010, Girard and Berthoz, 2005, Zhou et al., 2009]. The studies also provided evidence that saccades are generated through a parallel-distributed neural network, as shown in Fig. 2.1. The two sides of the symmetric network in Fig. 2.1 are known as the ipsilateral side and the contralateral side. The ipsilateral side exhibits coordinated activities in the initiation and control of the saccade in the right eye, while the contralateral side simultaneously synapses with the ipsilateral side to generate a saccade in the left eye. Each neuron in the parallel-distributed network fires in response to other neurons to stimulate the final motoneurons on both sides of the network to execute a binocular saccade. The neural populations on each side of the midline excite and inhibit one another sequentially to ensure that this coactivation leads to the coordination of movement between the eyes.

In the context of the neuroanatomical connectivity structure in Fig. 2.1, the saccade neural network includes neuron populations to imitate the behavior of actual neuronal populations in the initiation, control, and termination of the saccadic burst generator. Neural coordinated activities of the SC and the fastigial nucleus (FN) of the cerebellum are identified as the saccade initiator and terminator, respectively. The interactions between SC and basal ganglia, as well as activity of the key cortical areas in the flow of cortical visual processing, have their own implications in the saccadic premotor dynamics. We focus here on the saccade execution by the premotor burst generator neural sites in the midbrain. There are three key types of set of neurons that mimic the neural activates in a downstream pathway from the SC to the neuromuscular interface. First, the saccadic burst neurons elicit phasic movement command that is proportional to the saccade velocity. Second, the tonic neurons transform the phasic command to a tonic command by their internal integration state, thus releasing tonic spikes in relation with the amplitude of the saccade. The third types are motoneurons that aggregate the phasic and tonic commands into a final innervation signal. This signal first interacts with the lateral and medial rectus muscles to orient the eyeball to the saccade target (phasic component), and subsequently holds it in a static manner against the elastic restoring forces (tonic component). A basic description of the properties of the major neural sites involved in the execution of a saccade provides the basis for developing quantitative computational models of the neural network.

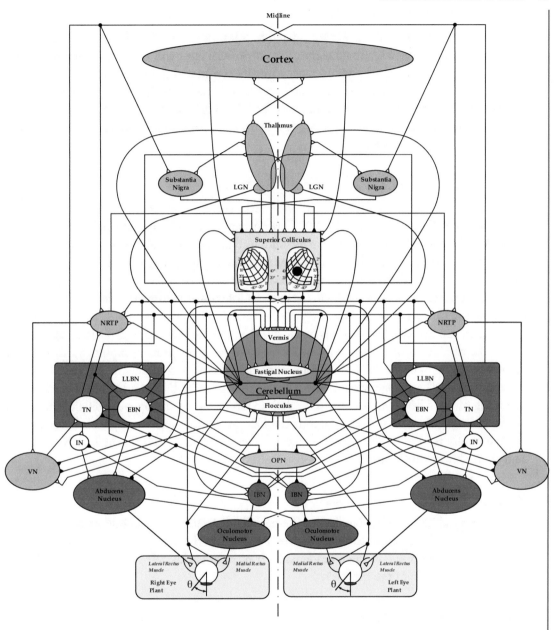

Figure 2.1: The parallel-distributed neural network for generation of a 20° conjugate goal-directed horizontal saccade in both eyes. Excitatory and inhibitory inputs are shown with white and black triangles at the postsynaptic neurons, respectively. This network is an updated network of that proposed by Enderle and Zhou [2010] such that IN mediates between TN and Abducens Nucleus (AN). In addition, the IN is inhibited by the IBN on each side.

2.2.1 SUPERIOR COLLICULUS

The SC initiates the saccade and is considered to translate visual stimuli to motor commands. It includes two important functional regions: the superficial layer and the deep layers [Enderle and Zhou, 2010]. The superficial layer is conventionally considered the visual layer that receives the information from the retina and the visual cortex. The deep layers, however, are involved with generation of the desired efferent commands for initiating saccades. It should be noted that the deep layers cause a high-frequency firing that starts 18–20 ms before a saccade, and ends almost when the saccade is complete.

2.2.2 PREMOTOR NEURONS IN THE PPRF

The paramedian pontine reticular formation (PPRF) encompasses neurons that show dominantly increasing burst frequencies of up to 1,000 Hz during the saccade and remain inactive during the periods of fixation. The LLBN and the medium lead burst neuron (MLBN) are the two types of burst neurons in the PPRF. The LLBN forms an excitatory synapse to the IBN and an inhibitory synapse to the OPN. Although the biophysical properties of the LLBN are not strongly related to the saccadic characteristics observed in the data, its functionality (direction selectivity) is essential to the control of the saccades.

There are two types of neurons in the MLBN: the EBN and the IBN. The EBN serves as one of the vital excitatory inputs for the saccade controller. The primary inputs to this neuron are the excitatory input of the SC and the inhibitory input from the contralateral IBN and OPN. This neuron forms excitatory synapses to the TN and the AN. The IBN, on the other hand, controls the firing of the EBN as well as the TN, both of which are on the opposite side of the network to the corresponding IBN. It also inhibits the ON and the IN ipsilaterally. This neuron receives excitatory inputs from the FN of the cerebellum contralaterally and from the LLBN ipsilaterally, and an inhibitory input from the OPN.

2.2.3 OMNIPAUSE NEURON

The OPN inhibits the MLBNs during the periods of fixation, and is inhibited by the LLBN during the saccade. It stops firing about 10–12 ms before the saccade starts, and resumes firing approximately 10 ms before the saccade ends. It receives exclusively inhibitory inputs from the LLBN on either side of the network.

2.2.4 TONIC NEURON

The TN is responsible for keeping the rectus eye muscles steady once the saccade completes. This neuron receives excitatory input from the corresponding EBN and inhibitory input from the opposite IBN. During saccades, the tonic neuron remains silent until the saccade ends. At this point, the tonic neuron generates a signal of variable frequency, depending on how far the eye has moved from its initial position. In particular, the tonic neuron functions as an integrator generating an action potential train whose frequency is directly proportional to the integrated EBN signal.

2.2.5   INTERNEURON

Many excitatory and inhibitory INs in the central nervous system stimulate and control motoneurons. The cerebellum aggregates most of these INs whose functionality depends on the anatomical aspects and properties of their membranes. The IN receives the excitatory and inhibitory inputs from the corresponding TN and IBN, respectively. It consecutively provides the step component to the agonist and antagonist neural controllers.

2.2.6   ABDUCENS NUCLEUS

The burst discharge in the motoneurons resembles a delayed EBN burst signal and is responsible for movement of the eyes conjugately. In motoneurons, the end structure of the axon is connected firmly to the muscle membrane. The AN drives the lateral rectus eye muscle, while also firing in synchrony with the ON from the opposite side. It is excited by the EBN during the saccade and by the IN once the saccade is completed. The IBN on the opposite side inhibits this neural population during the periods of fixation.

2.2.7   OCULOMOTOR NUCLEUS

The ON is solely responsible for the stimulation of the medial rectus eye muscle. This nucleus receives excitatory input from the opposite AN, and inhibitory input from the corresponding IBN.

2.2.8   CEREBELLUM

The cerebellum functions as a time-optimal gating element by using three active sites, namely, the cerebellar vermis (CV), the fastigial nucleus (FN), and the flocculus during the saccade [Enderle and Zhou, 2010]. The CV retains the current position of the eye by registering the information on the proprioceptors in the oculomotor muscles and an internal eye position reference. The CV also keeps track of the dynamic motor error used to control the saccade amplitude in connection with the nucleus reticularis tegmenti pontis (NRTP) and the SC. The FN is stimulated by the SC and projects ipsilaterally and contralaterally to the LLBN, IBN, and the EBN on the opposite side of the network. The contralateral FN starts bursting 20 ms before the saccade, while the ipsilateral FN undergoes a pause in firing and discharges with a burst slightly before the saccade completion. The third site, the flocculus, increases the time constant of the neural integrator for saccades with starting locations dissimilar to the primary position. By virtue of the physiological evidence, the cerebellum is responsible for terminating a saccade precisely, with respect to the primary position of the eye in the orbit [Enderle and Zhou, 2010].

2.3   FIRING CHARACTERISTICS OF EACH TYPE OF NEURON

The saccade generator investigated in this work is built upon the existing research [Enderle, 1994, Enderle and Engelken, 1995, Enderle, 2002, Enderle and Zhou, 2010, Enderle and Bronzino, 2011, Zhou et al., 2009]. Monkey saccades are categorized into two different modes of operation: small (ranging from 3° to 8°) and large (above 8°) [Enderle and Zhou, 2010]. The differentiation between these two modes has been governed by the fact that when the saccade size increases, more active neurons are firing synchronously to form the agonist neural input for small saccades. For large saccades, however, the number of active neurons firing maximally remains unchanged, consistent with the time-optimal controller described by Enderle and Wolfe [1987]. The model is first-order time-optimal; that is, it does not depend on the firing rate of the neurons to determine the saccade magnitude. We next describe necessary dynamical features of the proposed saccade neural network that affect the control of the saccades.

2.3.1   NEURAL ACTIVITY

The structure of the saccade neural network leverages a neural coding so that burst duration is transformed into saccade amplitude under the time-optimal condition. Such coding manifests activities—including the onset of burst firing before saccade, peak firing rate, and end of firing with respect to the saccade termination—for each neuron on the basis of the physiological evidence. These characteristics are provided for the neural sites as a framework for our simulations [Enderle and Zhou, 2010]. Table 2.1 summarizes the activities in initiation, control, and termination of the burst firing through the neural network, generating a saccade in the right eye. It is worthy of note that the agonist and antagonist tonic firing is governed by the ipsilateral IN activity under the tonic firing operation mode [Faghih et al., 2012].

2.3.2   BURST DISCHARGE MECHANISM

The firing rate trajectories of a medium lead burst neuron of monkey data for saccades of 4°, 8°, 12°, 16°, and 20° are provided [Enderle and Sierra, 2013]. It is explained that such trajectories are in agreement with the data published in the literature. The representation of the trajectories in Fig. 1.12 aids in comprehending the foundations of the first-order time-optimal neural controller. As demonstrated, the entire active agonist neurons fires maximally during the pulse interval of the saccade. For small saccades, the controller is constrained by a required minimum duration of the agonist pulse. Knowing this, the saccade magnitude depends on the number of active neurons, firing maximally, according to the physiological evidence [Enderle and Zhou, 2010]. As indicated in Section 1.6, the number of active neurons is the only parameter that varies in the MFM among different saccades in an adaptive control strategy of the oculomotor plant. It is found that adjusting this parameter provides significant analytical convenience in controlling the small saccades, as opposed to changing the firing rate of all active neurons as a function of saccade magnitude.

Table 2.1: Firing activity of neural sites during an ipsilateral saccade

Neural Site	Burst Onset Before Saccade (ms)	Peak Firing Rate (Hz)	Burst End with Respect to Saccade End
Contralateral SC	20–25	800–1000	Almost the same
Ipsilateral LLBN	20	800–1000	Almost the same
OPN	6–10	150–200 (before and after)	Almost the same
Ipsilateral EBN	6–8	600–1000	~10 ms before
Ipsilateral IBN	6–8	600–800	~10 ms before
Ipsilateral TN/IN	5	Tonic firing (before and after)	Resumes tonic firing when saccade ends
Ipsilateral AN	5	400–800	~5 ms before
Ipsilateral FN	20	Pause during saccade and a burst of 200 Hz near the end of the saccade	Pause ends with burst ~10 ms before saccade ends; resumes tonic firing ~10 ms after saccade ends
Contralateral FN	20	200	Pulse ends with pause ~10 ms before saccade ends; resumes tonic firing ~10 ms after saccade ends
Ipsilateral CV	20–25	600–800	~25 ms before
Ipsilateral NRTP	20–25	800–1000	Almost the same
Ipsilateral Substantia Nigra	40	40–100	Resumes firing ~40–150 ms after saccade ends

For the large saccades, however, the results establish that the duration of the agonist pulse is the dominant factor that determines the saccade magnitude. Such duration varies noticeably among the large saccades shown in Fig. 1.12. The FN in the cerebellum records the duration of the agonist pulse and the number of active neurons in arranging the end of the saccade.

Since motoneurons receive excitatory input from the ipsilateral EBN, the burst discharge in them during a saccade is adequately similar to the EBN bursting. Such burst discharge in the motoneurons is responsible for the movement of the rectus muscles during a saccade. The firing rate trajectory of the EBN is of prime importance in control of such a saccade. The proposed EBN model by Enderle and Zhou [2010] showed a constant plateau of bursting during the second portion of the burst before the decay occurs. We model the EBN firing rate by applying the firing rate trajectory in which a slow linear reduction in firing rate is assumed [Enderle and Zhou, 2010, Gancarz and Grossberg, 1998].

The existence of the SC excitation of the PPRF has been verified anatomically in the monkey. In particular, Keller et al. conducted electrophysiological experiments that attested the evidence of the direct projection from the SC to the LLBN [2000]. However, the evidence of the direct projection from the SC to the EBN was not confirmed. In contradiction to this study, there is sufficient evidence by merits of anatomical investigations supporting that this latter projection exists [Moschovakis et al., 1996, Olivier et al., 1993, Stanton et al., 1988]. The stimulation of the deeper layers of the SC in the monkey tended to EBN activity with triple-pulse stimuli [Raybourn and Keller, 1977]. The differences in saccade amplitude, duration, velocity, and the stability of the Excitatory Burst Neuron (EBN) in a monkey were examined [Hu et al., 2007]. We consider a gradual descent trajectory for the contralateral SC and FN stimulation of the ipsilateral LLBN, as shown in Fig. 2.2. These trajectories accord with those of the different simulations in examining the effects of several depolarizing stimulus currents in the EBN axon [Enderle and Zhou, 2010]. It should be emphasized at this point how the SC contributes to the optimal control of the saccades by driving the LLBN. The neural activity in the SC is arranged into movement fields that are related to the direction and saccade amplitude [Zhou et al., 2009]. The movement fields within the SC are indicators of the number of neurons firing for different small and large saccades (see locus of points on a detailed view of the SC retinotopic mapping in Fig. 2.14 in Enderle and Zhou [2010].

Neurons active in the SC in Fig. 2.1 are shown with the dark circle, representing the locus of points for a desired 20° saccade. Enderle and Zhou reported that active neurons in the deep layers of the SC generate a sporadic high-frequency burst of activity that varies with time, initiating 18–20ms before a saccade and ending sometime toward the end of the saccade [2010]. However, the exact timing for the end of the SC firing happens quite randomly and can be either before or after the saccade ends. It is implied that the number of cells firing in the LLBN is determined by the number of cells firing in the SC, as long as there is a feedback error maintained by the cerebellar vermis. The number of the OPN cells firing after inhibition from the LLBN determines, in turn,

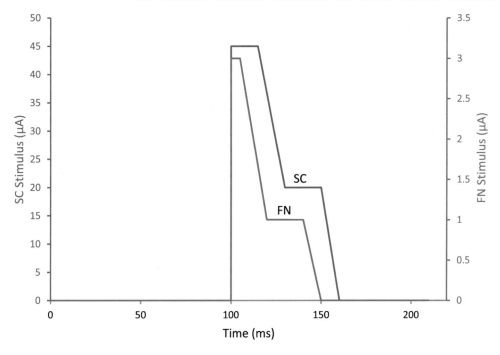

Figure 2.2: The current stimulation trajectories of the ipsilateral LLBN. The current amplitude for the contralateral SC and contralateral FN is chosen based on the burst properties for these two neural sites provided in Table 2.1. As for the contralateral FN stimulus current, a step current of 3μA is applied at 100 ms, shortly after which a linear decrease in stimulus to 1μA appears. Subsequently, another step stimulation continues until 140 ms, when a linear reduction occurs until the current is removed at 150 ms.

how many EBN cells are released from inhibition. In consequence, the number of EBN cells firing determines the number of motoneurons driving the eyes to their destination.

2.3.3 SEQUENCE OF NEURAL FIRING

The saccade completion involves the evolution of some events in an orderly sequence in the neural sites. Such neural sites are shown in Fig. 2.3 via a functional block diagram [Enderle and Zhou, 2010]. The output of each block indicates the firing pattern at each neural site manifested during the saccade: saccade begins at time zero, and T represents the saccade termination. The negative time for each neural site refers to the onset of the burst before the saccade (see Table 2.1). The neural activity within each block is represented as pulses and/or steps, consistent with the described burst discharge mechanism, to reflect the neural operation as timing gates. Ultimately,

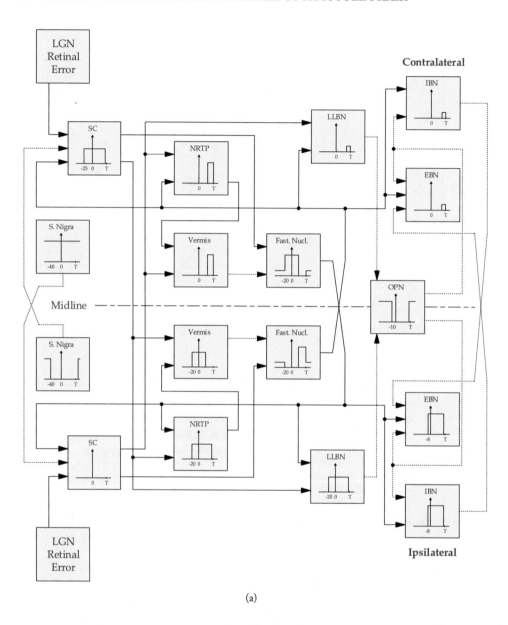

(a)

Figure 2.3: A functional block diagram of the saccade generator model [Enderle and Zhou, 2010]. Solid lines are excitatory and dashed lines are inhibitory. Each block represents the neural activity of the corresponding neural site as indicated in Table 2.1. (a) Neural pathways from the formation of the lateral geniculate nucleus (LGN) retinal error to the MLBN. *(Continues.)*

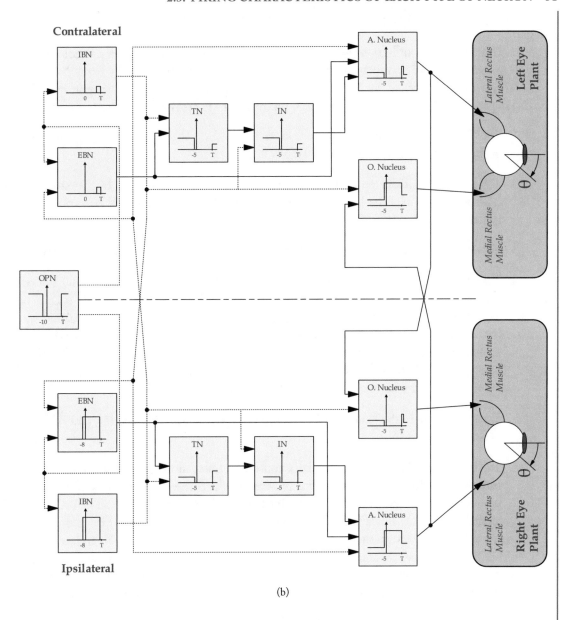

(b)

Figure 2.3: *(Continued.)* A functional block diagram of the saccade generator model [Enderle and Zhou, 2010]. Solid lines are excitatory and dashed lines are inhibitory. Each block represents the neural activity of the corresponding neural site as indicated in Table 2.1. (b) Neural pathways from the MLBN to the rectus muscles in both eyes. The IN directs the synapses between the TN and motoneurons.

motoneurons innervate rectus muscles in both eyes at the end interaction level of the block diagram.

The following outlines eight steps required to implement the saccade control strategy in the context of Fig. 2.3. It represents the sequence of events accounted for by Enderle and Zhou [2010], with modifications made in steps 4–7. The function of local neural integrators (TN and IN) in providing the step component to the motoneurons is provided as such modifications.

1. The deep layers of the SC initiate a saccade based on the distance between the current position of the eye and the desired target.

2. The ipsilateral LLBN and EBN are stimulated by the contralateral SC burst cells. The LLBN then inhibits the tonic firing of the OPN. The contralateral FN also stimulates the ipsilateral LLBN and EBN.

3. When the OPN ceases firing, the MLBN (EBN and IBN) is released from inhibition.

4. The ipsilateral IBN is stimulated by the ipsilateral LLBN and the contralateral FN of the cerebellum. When released from inhibition, the ipsilateral EBN responds with a post-inhibitory rebound burst for a brief period of time. The EBN, when stimulated by the contralateral FN (and perhaps the SC), enables a special membrane property that causes a high-frequency burst that decays slowly until inhibited by the contralateral IBN. The burst-firing activity of EBN is integrated through the connection with the TN. The IN follows closely the same integration mechanism as that of the TN.

5. The burst firing in the ipsilateral IBN inhibits the contralateral EBN, IN, and AN, as well as the ipsilateral ON.

6. The burst firing in the ipsilateral EBN causes the burst in the ipsilateral AN, which then stimulates the ipsilateral lateral rectus muscle and the contralateral ON. With the stimulation of the lateral rectus muscle by the ipsilateral AN, and the inhibition of the ipsilateral medial rectus muscle via the ON, a saccade occurs in the right eye. Simultaneously, the contralateral medial rectus muscle is stimulated by the contralateral ON, and, with the inhibition of the contralateral lateral rectus muscle via the AN, a saccade occurs in the left eye. Hence, the eyes move conjugately under the control of a single drive center. During the fixation periods, the INs provide the steady-state tensions required to keep the eyes at the desired destination.

7. At the termination time, the cerebellar vermis, operating through the Purkinje cells, inhibits the contralateral FN and stimulates the ipsilateral FN. Some of the stimulation of the ipsilateral LLBN and IBN is lost because of the inhibition of the contralateral FN. The ipsilateral FN stimulates the contralateral LLBN, EBN, and IBN. The contralateral EBN then stimulates the contralateral AN. The contralateral IBN then inhibits the ipsilateral

EBN, TN, and AN, and the contralateral ON. This inhibition removes the stimulus to the agonist muscle.

8. The ipsilateral FN stimulation of the contralateral EBN allows for modest bursting in the contralateral EBN. This activity then stimulates the contralateral AN and the ipsilateral ON. Once the SC ceases firing, the stimulus to the LLBN stops, allowing the resumption of OPN firing that inhibits the ipsilateral and contralateral MLBN, hence terminating the saccade.

The advances in computational neural modeling have supplied us with abundant information at different structural scales, such as the biophysical [Ghosh-Dastidar and Adeli, 2007, 2009, Mohemmed et al., 2012], the circuit [Enderle and Zhou, 2010, Rosselló et al., 2009], and the systems levels [Ramanathan et al., 2012]. The following includes our modeling of the premotor and motor neurons at the circuit level. We introduce a neural circuit model that can be parameterized to match the described firing characteristics of each type of neuron.

2.4 NEURAL MODELING

A typical neuron embodies four major components: cell body, dendrites, axon, and presynaptic terminals, as shown in Fig. 2.4. The neural cell body encompasses the nucleus, as is true of other cells. Dendrites act as the synaptic inputs for the preceding excitatory and inhibitory neurons. Upon this stimulation of the neuron at its dendrites, the permeability of the cell's plasma membrane to sodium intensifies, and an action potential moves from the dendrite to the axon [Enderle and Bronzino, 2011]. The transmission of an action potential along the axon is facilitated by means of nodes of Ranvier in the myelin sheath. At the end of each axon there are presynaptic terminals, from which the neurotransmitters diffuse across the synaptic cleft.

A complete understanding of the properties of a membrane by means of standard biophysics, biochemistry, and electronic models of the neuron will lead to a better analysis of membrane potential response. A neuron circuit model is desired to quantify the saccade-related neural activity, thus reflecting the physiology linked to the dendrite, cell body, axon, and presynaptic terminal of each neuron. Such a model is sketched in this section, together with the description of its modifications, required to populate the neural network for the control of saccades. The saccade neural network includes eight neuron populations at premotor and motor levels as seen in Fig. 2.1:

1. long lead burst neuron (LLBN);

2. omnipause neuron (OPN);

3. excitatory burst neuron (EBN);

4. inhibitory burst neuron (IBN);

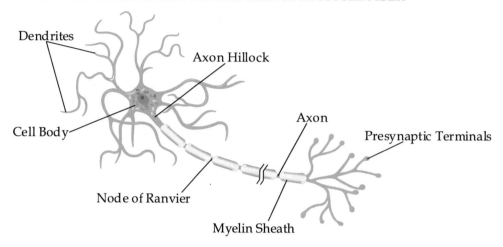

Figure 2.4: A schematic presentation of the different components of a neuron [Enderle and Bronzino, 2011].

5. tonic neuron (TN);

6. interneuron (IN);

7. abducens nucleus (AN);

8. oculomotor nucleus (ON).

The saccade circuitry underlies the dynamics of the above eight distinct neurons, each of which contributes to the control mechanism of the saccade. Except for the OPN, the proposed parallel-distributed neural network accommodates two of each of the other neurons in the network. The dendrite model delineated below is adjustable to the stimulation mechanism of all eight neurons. The axon model for all spiking neurons, except the EBN and OPN, adheres to the Hodgkin—Huxley (HH) model. The EBN and OPN are neurons that fire automatically when released from inhibition—these neurons are modeled using a modified HH model [Enderle and Zhou, 2010]. The TN integrates its input and is modeled with a FitzHugh—Nagumo (FHN) model under the tonic bursting mode [Faghih et al., 2012]. The presynaptic terminal elicits a pulse train stimulus whose amplitude depends on the membrane characteristics of the postsynaptic neuron.

2.4.1 DENDRITE MODEL

The dendrite is partitioned into a number of membrane compartments, each of which has a predetermined length and diameter. Each compartment in the dendrite has three passive electrical characteristics: electromotive force (emf), resistance, and capacitance, as shown in Fig. 2.5. Axial resistance is used to connect the dendrite to the axon.

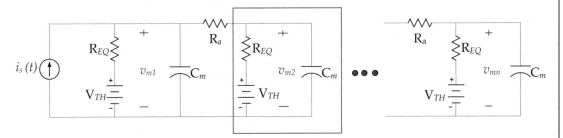

Figure 2.5: The dendrite circuit model with n passive compartments: $i_s(t)$ models the stimulus current from the adjacent neurons to the dendrite. Each compartment has membrane electromotive, resistive, and capacitive properties—V_{TH}, R_{EQ}, and C_m in the second compartment are noted. The batteries in the circuit, V_{TH}, are the Thévenin equivalent potential of all the ion channels. The axial resistance R_a connects each compartment to the adjacent ones (remains unchanged among the neurons). Appropriate values for the membrane resistance and capacitance of the dendrite model are found to match physiological evidence for each neuron.

The presynaptic input to the dendrite is modeled as a pulse train current source (i_s). The node equation for the first dendrite compartment is

$$C_m \frac{d v_{m1}}{dt} + \frac{v_{m1} - V_{TH}}{R_{EQ}} + \frac{v_{m1} - v_{m2}}{R_a} = i_s, \tag{2.1}$$

where v_{m1} is the membrane potential of the first compartment, and v_{m2} is the membrane potential of the second compartment. The membrane resistance R_{EQ}, capacitance C_m, and emf V_{TH} characterize each compartment. R_a is the axial resistance.

For all intermediate dendrite compartments there are two inputs: the input from the previous compartment's membrane potential and the input from the next compartment's membrane potential. The node equation for the second compartment is

$$C_m \frac{d v_{m2}}{dt} + \frac{v_{m2} - V_{TH}}{R_{EQ}} + \frac{v_{m2} - v_{m1}}{R_a} + \frac{v_{m2} - v_{m3}}{R_a} = 0, \tag{2.2}$$

where v_{m3} is the membrane potential of the third compartment.

The last dendrite compartment receives just one input from its preceding compartment. The corresponding node equation is

$$C_m \frac{d v_{mn}}{dt} + \frac{v_{mn} - V_{TH}}{R_{EQ}} + \frac{v_{mn} - v_{m(n-1)}}{R_a} = 0, \tag{2.3}$$

where the membrane potential v_{mn} is related to the preceding compartment's membrane potential ($v_{m(n-1)}$) through the axial resistance R_a.

The dendrite model of each neuron is accomplished by fine-tuning the parametric capacitance and resistance properties of the above-defined dendrite model. This parametric adaptation allows for the accommodation of the synaptic transmission in the neural network, as required to stimulate each postsynaptic neuron. Each neuron's dendrite rise time constant determines the delay to emulate the postsynaptic potential propagation along the dendrite. This is consistent with the onset of firing with respect to the saccade onset provided in Table 2.1. This time constant is inferentially determined, and verified with a circuit simulation suite for each neuron, such that its membrane potential reaches the threshold in synchrony with the sequential burst firing indicated previously. For instance, from the EBN's dendrite Thévenin equivalent circuit, nearly five time constants provides the necessary time delay between the OPN's end of firing and the EBN's onset of firing. Table 2.2 includes the membrane resistance and capacitance of the dendrite compartments for each neuron.

Table 2.2: Parametric realization of eight distinct neurons in terms of dendritic, axonal, and synaptic behaviors in the proposed neural circuitry

	Dendrite			Axon	Synapse
Neuron	Capacitor (μF)	Resistor (kΩ)	Firing threshold voltage (mV)	Coefficient	Pulse amplitude (μA)
LLBN	0.5	3.75	−45	18,000	20
OPN	1.0	6.3	−60	1,800	45
EBN	0.45	3.1	−60	35,000	75
IBN	0.35	4.5	−45	15,000	65
AN	0.35	5.5	−45	17,000	55
ON	0.45	4.0	−45	17,000	55
TN	0.35	4.5	NA	NA	10
IN	0.4	4.5	NA	NA	10

The initial condition (state) of the capacitor is set to V_{TH}. Computational efficiency accrues when the minimum number of compartments in the dendrite model is required. We chose to include 14 compartments in the dendrite to achieve the desired membrane properties in each type of neuron. For example, the EBN dendritic membrane potential across the first, second, third, and last compartments is illustrated in Fig. 2.6. The farther the compartment is along the dendrite, the smoother its potential response to the pulse train current source. The last compartment of the post-synaptic dendrites (cell body) leads the signal flow to the axon—the site of action potential generation.

2.4.2   AXON MODEL

Quite a few circuit models can be considered to reproduce the electrical properties of an axon in simulation of SNNs. The choice has to set forth a compromise between several factors, such as physiological realism, computational cost, complexity, accuracy, and scalability. Roy presented sodium and potassium conductance circuits of the field effect transistors (FETs) that precisely evoked the time dependency of each ion channel [1972]. The circuitry attains a high degree of physiological pragmatism, but it remains intricate in modification to match the firing specifications of a network of neurons. The Hodgkin—Huxley (HH) model of the axon serves as the basis for the neurons modeled here—only the EBN and the OPN are based on a modified HH model. As elaborated later, this modification leads these neurons to fire automatically at high rates after being released from inhibition, given minor stimulation. The HH model is developed to describe the membrane potential at the axon hillock caused by conductance changes [Enderle and Zhou, 2010]. The circuit diagram of an unmyelinated portion of squid giant axon is illustrated in Fig. 2.7. According to this circuit model, the sodium and potassium conductances are configured in parallel with a capacitor and in series with a battery—Nernst potential for each ion channel. The node equation that expresses the membrane potential V_m as a function of stimulus current I_m from the dendrite and voltage-dependent conductances of the sodium and potassium channels is

$$\bar{g}_K \, N^4 \, (V_m \, - \, E_K) + \bar{g}_{Na} \, M^3 \, H \, (V_m \, - \, E_{Na})$$
$$+ \frac{(V_m \, - \, E_l)}{R_l} \, + \, C_m \, \frac{d V_m}{dt} \, = \, I_m, \tag{2.4}$$

where

$$\frac{dN}{dt} = \alpha_N \, (1 \, - \, N) \, - \, \beta_N \, N,$$
$$\frac{dM}{dt} = \alpha_M \, (1 \, - \, M) \, - \, \beta_M \, M,$$
$$\frac{dH}{dt} = \alpha_H \, (1 \, - \, H) \, - \, \beta_H \, H,$$
$$\bar{g}_K = 36 \, \times \, 10^{-3} \, S, \qquad \bar{g}_{Na} \, = \, 120 \, \times \, 10^{-3} \, S.$$

The coefficients in the above first-order system of differential equations are related exponentially

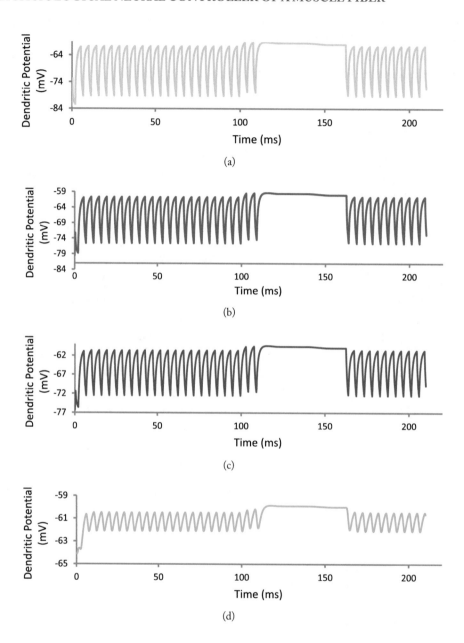

Figure 2.6: The EBN dendritic membrane potential across the different compartments. (a) first compartment, (b) second compartment, (c) third compartment, and (d) last compartment. The membrane parameter values are: $V_{TH} = -60$ mV, $C_m = 0.45\,\mu$F, $R_{EQ} = 3.1$ kΩ, and $R_a = 100\,\Omega$.

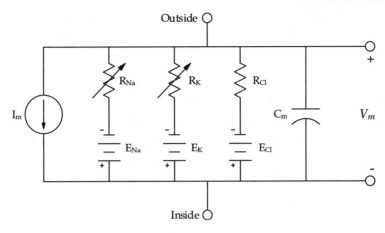

Figure 2.7: The circuit model of an unmyelinated portion of squid giant axon [Enderle and Zhou, 2010]. The variable active-gate resistances for Na^+ and K^+ are given by $R_K = 1/\bar{g}_K N^4$ and $R_{Na} = 1/\bar{g}_{Na} M^3 H$ respectively. The passive gates are modeled by a leakage channel with resistance, $R_l = 3.33\,k\Omega$. The battery is the Nernst potential for each ion: $E_l = 49.4\,V$, $E_{Na} = 55\,V$, and $E_K = 72\,V$.

to the membrane potential V_m, i.e.,

$$\alpha_N = 0.01 \times \frac{V + 10}{e^{\left(\frac{V + 10}{10}\right)} - 1}\ ms^{-1}, \qquad \beta_N = 0.125\, e^{\left(\frac{V}{80}\right)}\ ms^{-1},$$

$$\alpha_M = 0.1 \times \frac{V + 25}{e^{\left(\frac{V + 25}{10}\right)} - 1}\ ms^{-1}, \qquad \beta_M = 4\, e^{\left(\frac{V}{18}\right)}\ ms^{-1}, \qquad (2.5)$$

$$\alpha_H = 0.07\, e^{\left(\frac{V}{20}\right)}\ ms^{-1}, \qquad \beta_H = \frac{1}{e^{\left(\frac{V + 30}{10}\right)} + 1}\ ms^{-1},$$

$$V = V_{rp} - V_m\ mV$$

where the resting potential V_{rp} is $-60\,mV$.

The neural firing rate of all the bursting neurons has been adjusted to meet the peak firing rate requirement in Table 2.1. This adjustment intends for each neuron to contribute to the generation of the saccade by mimicking the required physiological properties [Enderle and Zhou, 2010]. To this end, the right-hand side of the N, M, and H differential expressions in Eq. (2.4) is multiplied by appropriate coefficients to achieve the desired peak firing rates. For instance, the required coefficient for the EBN is 35,000, therefore it presents a peak firing rate at 1,000 Hz.

It should be pointed out that the above equations of the basic HH model of the axon have been used for all the bursting neurons, except for the EBN and the OPN. For these latter neurons, the modified HH model is used to change the threshold voltage from -45 mV to -60 mV. Enderle and Zhou [2010] illustrated experiments in which this variation caused EBN to fire autonomously without the existence of any excitatory stimulus. From their description of the dominant effect of the sodium channel current on the changes in the threshold voltage at the beginning of the action potential, the threshold voltage in the EBN axon model is changed by modifying the α_M equation to

$$\alpha_M = 0.1 \times \frac{V + 10}{e^{\left(\frac{V + 10}{10}\right)} - 1} \, ms^{-1}. \tag{2.6}$$

The OPN axonal threshold voltage of firing has been adjusted following the same modification by the above equation. This alteration of the threshold voltage for the EBN and the OPN enables them to fire spontaneously without any significant depolarization from peripheral current stimuli. Table 2.2 lists the firing threshold voltage and the coefficient required to adjust the peak firing rate for each bursting neuron.

The axon transfers an action potential from the spike generator locus to the output end, the synapse. The transmission along the axon thus amounts to introducing a time delay, after which the action potential appears at the synapse.

2.4.3   SYNAPSE MODEL

When the action potential appears at the synapse, packets of neurotransmitter are released. This is modeled by excitatory or inhibitory pulse train stimuli to stimulate the dendrite of the postsynaptic neuron more realistically. Current-based synapse models offer significant analytical convenience when describing how a postsynaptic current pulse is triggered by an action potential in very large SNNs [Wong et al., 2012]. As these models disregard the voltage-dependent property of the postsynaptic currents, for the networks with both the interspike intervals and the burst onsets of the neurons uniformly distributed, they are preferred to the conductance-based synapse models.

In the current-based synapse models, the total synaptic current is the linear combination of all synaptic currents induced by individual spikes to each neuron. This partial induction of current is described by Wong et al. [2012], as a modified α-function, which follows for an inhibitory synapse

$$I_i^{syn}(t) = \sum_k I_{i,k}^{syn}(t) = \sum_k \alpha_{i,k}(t), \tag{2.7}$$

where for the kth spike in the spike chain to the neuron i, $I_{i,k}^{syn}(t) = \alpha_{i,k}(t)$. This function is termed as

$$
\alpha_{i,k}(t) = \begin{cases} 0 & t < t_{i,k} + d \\ G \dfrac{1}{\tau_c^d - \tau_c^r} \left(e^{-\frac{t - t_{i,k} - d}{\tau_c^d}} - e^{\frac{t - t_{i,k} - d}{\tau_c^r}} \right) & t \geq t_{i,k} + d, \end{cases} \tag{2.8}
$$

where G is a constant representing the strength of the synaptic current, and τ_c^d and τ_c^r denote the decay and rise time constants, respectively. $t_{i,k}$ is the onset of the kth spike to neuron i, and d represents the delay in synaptic transmission. It can be shown that for a sufficiently low rise time constant and a properly high decay time constant, the time course of the function $\alpha_{i,k}(t)$ above can be approximated by a rectangular pulse. In this limiting case, Eq. (2.8) reduces to

$$
\alpha_{i,k}(t) = \begin{cases} 0 & t < t_{i,k} + d \\ G'(\tau_c^d, \tau_c^r) & t \geq t_{i,k} + d, \end{cases} \tag{2.9}
$$

where G' is postulated as the amplitude of the rectangular pulse, to a very good approximation. It follows thus that the time dependency of the postsynaptic current pulses is the same for each incoming presynaptic spike. Thereby, the synaptic current during each firing interval of a neuron can be realized a priori and invoked from the memory only during the simulation. This feature offers a great deal of computational efficiency when dealing with the large-scale polysynaptic spiking neural networks.

In view of the release of the rectangular synaptic pulses from each neuron, the amplitude and width of these pulses are determined in simulation runs by tweaking them, to provide the desired postsynaptic behavior in the interconnected neurons. The width is constrained by the two points at which the action potential crosses a constant level of the axonal potential.

Figure 2.8 shows a number of action potentials and the synaptic current pulses of the EBN toward the end of the burst firing interval. The time delay between each action potential and the corresponding current pulse is apparent.

In addition to the transmission time delay along the axon, all chemical synapses introduce a small delay before the generation of postsynaptic potentials from an input excitatory or inhibitory pulse train. This delay accounts for the time required for the release of neurotransmitters and the time it takes for them to distribute through the synaptic cleft. This small synaptic delay was taken into effect by increasing the rise time constant of the subsequent postsynaptic dendritic compartments.

As indicated, the amplitude and width of synaptic current pulses for each neuron are uniquely chosen in order that the postsynaptic neurons exhibit the desired behavior. Table 2.2 includes such amplitude of the synaptic current pulses. This table summarizes all the differences (dendritic, axonal, and synaptic) among eight distinct neurons whose realization is important in the neural circuitry for time-optimal control of the saccade.

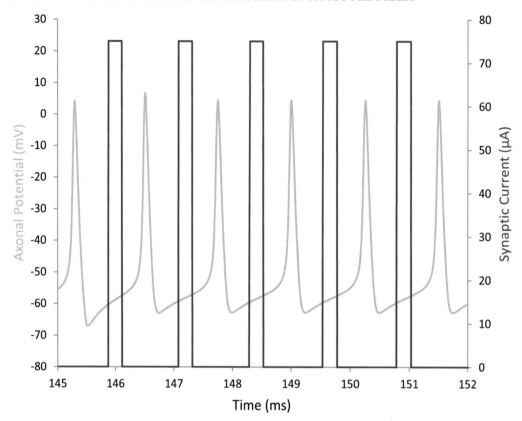

Figure 2.8: A train of action potentials and current pulses reflecting the synaptic transmission in the EBN. Each current pulse shows a time delay with respect to the corresponding action potential, due to the transmission delay along the axon.

2.5 TIME-OPTIMAL CONTROL OF A MUSCLE FIBER MODEL

A neural controller mechanism is required to relate the motoneuronal firing rates to the rectus muscles displacement. Through the examination of the spectral main sequence of saccades, it was found that saccades are not likely to be driven by a rectangular pulse bang-bang controller [Harwood et al., 1999]. In addition, it was determined that the neural input to the saccade system has a time-optimal pulse-step waveform, contradicting the pulse waveform control hypothesis [Enderle and Wolfe, 1987, 1988].

Continuing progression of the research indicated that for saccade generation in the right eye, an agonist pulse-slide-step controller stimulates the agonist muscle, and an antagonist pause-step controller inhibits the antagonist muscle [Enderle and Zhou, 2010, Zhou et al., 2009]. In the left eye, on the other hand, the antagonist pulse-slide-step controller stimulates the antagonist muscle, whereas the agonist muscle is inhibited by the agonist pause-step controller.

The time-optimal controller model is investigated herein to obtain the saccadic eye movement solution that drives the eyeball to its destination for different saccades. Such solution is characterized by the realization of two complimentary controllers; that is, the agonist controller model and the antagonist controller model. These models describe the dynamics by which the motoneuronal innervation signals are converted to the active-state tensions. The resulting active-state tensions are then used as inputs to the linear homeomorphic muscle fiber model (MFM), introduced in Chapter 1.

2.5.1   AGONIST CONTROLLER MODEL

The agonist controller is a first-order pulse-slide-step neural controller that describes the agonist active-state tension as the low-pass filtered neural stimulation signal. The expression of low-pass filtering of the neural innervation input to the agonist controller model is given in Eq. (1.40). The neural stimulation signal is the firing rate of the ipsilateral AN and that of the contralateral ON. Given in Eq. (1.43) is the expression for the agonist time constant, τ_{ag}. It is described by two step functions including the agonist activation time constant, τ_{gac}, and the deactivation time constant, τ_{gde}.

2.5.2   ANTAGONIST CONTROLLER MODEL

The antagonist muscle is unstimulated by a pause during the saccade, and remains fixed by a step input to keep the eyeball at its destination. To serve this purpose, a first-order pause-step neural controller is defined (Eqs. (1.42) and (1.44)). The neural stimulation signal to the controller is the firing rate of the ipsilateral ON and that of the contralateral AN. For the normal saccades with no post-inhibitory rebound burst firing in the antagonist motoneurons, we re-write Eq. (1.44) to

$$\tau_{ant} = \tau_{tde}\left(u(t - T_1) - u\left(t - T_3\right)\right) + \tau_{tac}u\left(t - T_3\right), \qquad (2.10)$$

where the antagonist time constant is described by two step functions, introducing the antagonist deactivation time constant, τ_{tde}, and the activation time constant, τ_{tac}. T_1 accounts for the latent period, and T_3 is the onset of the change to the step component necessary to keep the eyeball steady at its destination.

2.5.3   MUSCLE FIBER OCULOMOTOR MODEL

The oculomotor system is thought to be actuated by the command signals from the neural controllers. The anatomical and mathematical descriptions of the muscle fiber model (MFM) of muscle were the purpose of Chapter 1. The experiments with the MFM were illustrated for different combinations of columns of series of muscle fibers, load mass and active-state tension. As indicated, the significance of introducing a muscle fiber model is that it accommodates multiple neurons to drive the eyes to their destination. Accordingly, the effect of the number of active neurons in controlling the saccade magnitude can be investigated in an adaptive control paradigm of the oculomotor plant. In contrast to the whole muscle model, information about the muscle fibers is not aggregated into just a few parameters in the MFM. The rigorous analysis of the MFM including both the length-tension and the force-velocity characteristics indicates that this model agrees with the previous results from the whole muscle model (see Section 1.5).

Fig. 1.6 shows the investigated oculomotor plant with two parallel networks of the muscle fibers attached to the eyeball. Therein, the MFM of the agonist and antagonist rectus eye muscles are incorporated. The use of the state-variables approach facilitated the mathematical description of the MFM. The dynamics of the agonist MFM is governed by Eqs. (1.28)–(1.31). In addition, Eqs. (1.32)–(1.36) provide the dynamics of the antagonist MFM. Note that the agonist and agonist active-state tensions form the above neural controllers stand to be plugged into Eq. (1.30) and (1.34), respectively. Ultimately, the linear differential equation to solve for the time-optimal solution for corresponding change in length of eyeball arc is Eq. (1.37).

It is noteworthy that the neural stimulation analysis in this chapter differs from that of the previous chapter. The analytical solution for neural input to the MFM is achieved in Section 1.4. However, such neural input is not systematically provided herein, but it is attained by implementing an SNN to reproduce the electrophysiological burst properties of the motoneurons. Aside from this difference, no empirical parameters are involved herein other than the parameters of the oculomotor plant for monkey [Enderle and Zhou, 2010]. The simulation specifications and results follow.

2.6 NEURAL SYSTEM IMPLEMENTATION

Two small saccades (4° and 8°) and three large saccades (12°, 16°, and 20°) have been the focal point of our simulations of horizontal monkey saccades under the first-order time-optimal control strategy. All neural populations consisted of 14 dendrite compartments with membrane properties included in Table 2.2. The determination of the rise time constant for each neuron's dendrite plays a vital role in the integration of current pulses at the synapse. Analyses of the dendritic membrane potentials were performed with the NI Multisim circuit design suite, and the neural network was simulated in the MATLAB/Simulink software. The modular programming and test of each individual neuron were achieved to constitute our Simulink model of the system of neurons at the highest level of the hierarchy. This implementation, in particular, intends to determine if all the timing requirements are achieved for each module. If a module had yet to satisfy its dynamical features at any stage of the implementation, it was modified and resimulated. More specifications about this implementation follow.

2.6.1   SIMULINK PROGRAMMING

Programming schemes implementing the artificial neural networks (not as the one we described here) mostly use look-up tables to simulate the neurons. The look-up table for each neuron conveys the information about its connections, input weight values, transfer function, and output equation, to describe the entire neurodynamics. A physiologically based model of the neuron is modeled herein, for which a program simulates the underlying membrane differential equations. The main advantage of this neural system is that it offers memory efficiency in allocating the neural activity to each neuron. The information about the neural processing elements (merely the invoked file of parameters listed in Table 2.2) is stored for each neuron. This neural network programming eliminates the barrier of computational cost noticeably. The program is developed in modular structures, thus allowing for analyzing each module to verify whether or not it meets the desired dynamic performance.

The first step to create the block diagram program of any system is to obtain its quantitative mathematical models. For the linear time-invariant dynamic systems, the input-output relationship can be derived in the form of transfer functions. Within the Simulink's block-oriented structure, the transfer function blocks can then be arranged into block diagrams, which are capable of showing the system interconnections graphically. Block diagram representation for a program structure can be implemented in the form of functional modules. As a consequence, each module can be individually developed, tested, and debugged. Finally, when all of the modules meet the desired dynamic performance, they can be linked together to form the main, functioning program of the top-level system. The ode23t solver with variable-step time resolution of the simulator is used to exercise the real-time operation of the neuron model. The program stop time is 210 ms. In the following, we provide an illustration of this top-level system together with its key subsystems, which allows for a review of many concepts presented in this chapter.

The modular structure of the main program of the neural network is presented in Fig. 2.9. This figure demonstrates the program for the block diagram representations of the neural network shown in Fig. 2.3. The program steps through the execution of its modules in an orderly sequence, such that a series of handshake events occurs, as presented in Section 2.3. The SC and FN modules thus have the highest order of execution, the LLBN the next highest, and so on, down to the motoneurons. The time a module takes to execute the synaptic stimuli agrees with the timing properties of the burst firing listed in Table 2.1.

There are a total of 415 blocks (input control ports, processing elements, and output ports) in the main program. Shown in Fig. 2.10 is the block diagram of modules for the EBN. Each module (dendrite, axon, and synapse) depicts a subsystem that is developed separately from the main program. We now engage in a closer look at a number of major properties of the program for each module.

Figure 2.11a shows the program implementing the EBN first dendrite compartment. Recall that Eq. (2.1) is the differential equation that describes the dynamics of the membrane potential of the first compartment. When each synaptic signal flows to the postsynaptic neuron, this signal

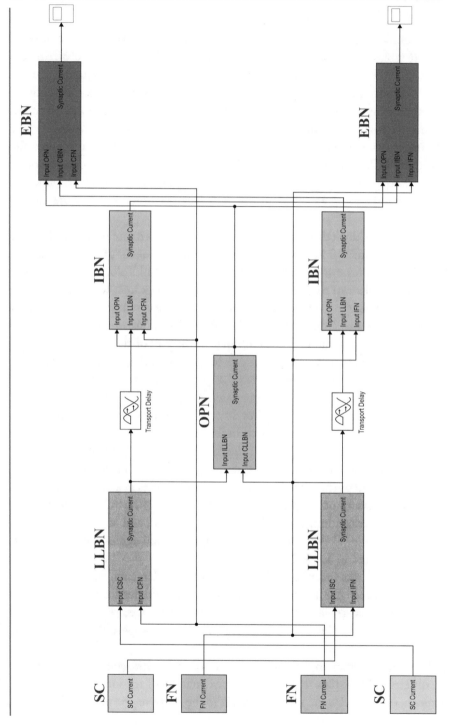

Figure 2.9: The top-level program of the saccade neural network. (a) From the FN and SC modules to the MLBN modules. (*Continues.*)

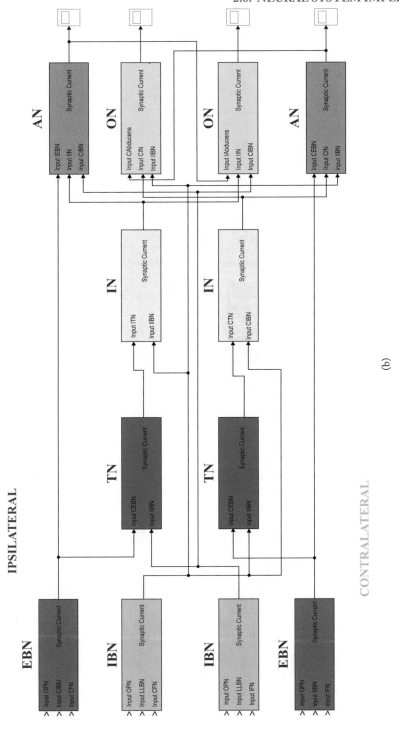

Figure 2.9: *(Continued.)* The top-level program of the saccade neural network. (b) From the MLBN modules to the motoneurons modules.

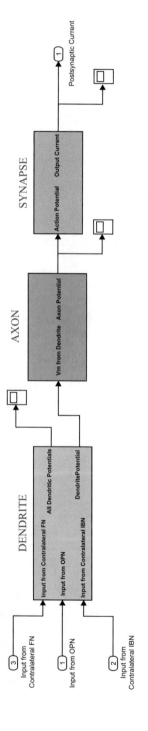

Figure 2.10:   Modules of the neural network program for the EBN.

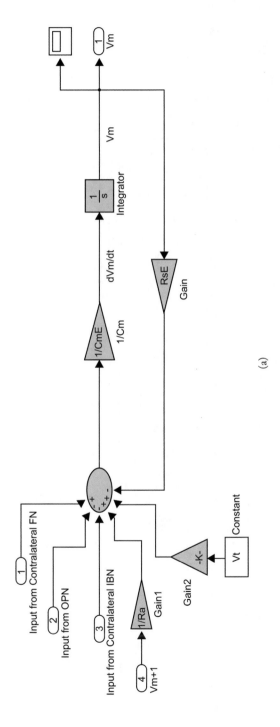

(a)

Figure 2.11: The modules of the program for the EBN dendritic compartments. (a) The initial compartment. (*Continues.*)

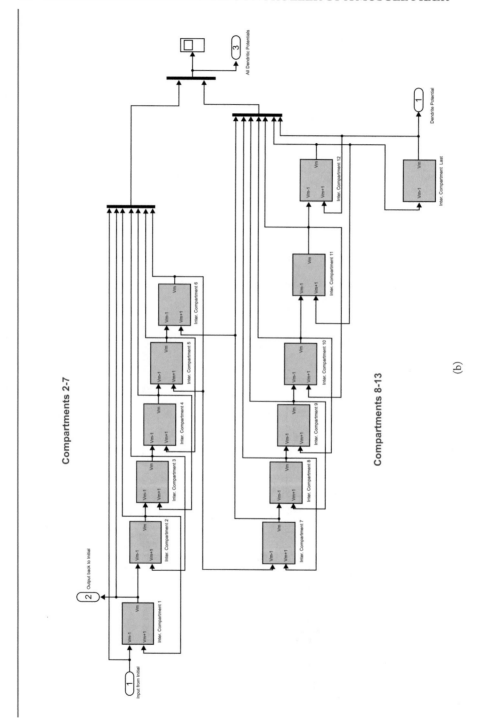

(b)

Figure 2.11: *(Continued.)* The modules of the program for the EBN dendritic compartments. (b) Blocks of the intermediate compartments (note the communication layout of the compartments). *(Continues.)*

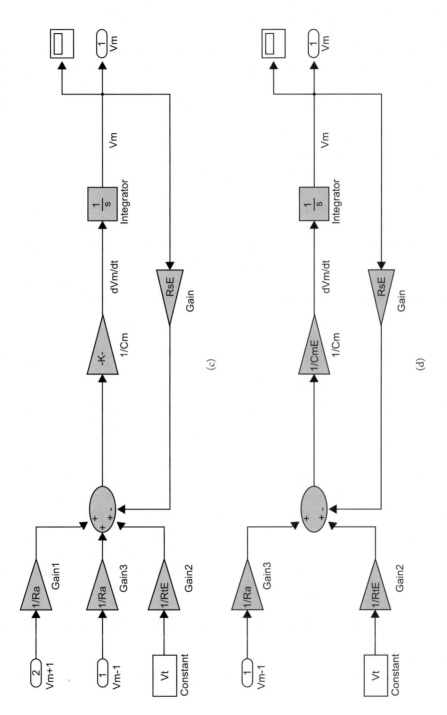

Figure 2.11: *(Continued.)* The modules of the program for the EBN dendritic compartments. (c) The fourth compartment block units. (d) The last compartment block units.

is superimposed with the others to produce a pulse train current source. In the view of this, it is clear that the summation block interfaces the postsynaptic signals with the dendrite compartments. The sign of the input from the synapses to the EBN depends on whether the synapse is excitatory or inhibitory. The output of the integrator is the membrane potential (one of the system's state variables). The block diagram representations of interconnections of the EBN dendrite compartments are depicted in Fig. 2.11a. Evidently, each intermediate compartment receives one input from the previous compartment's membrane potential and the other from the next compartment's membrane potential. Figure 2.11a displays the program for the EBN fourth dendrite compartment. The underlying differential equation is Eq. (2.2) that expresses the dynamics of the membrane potential of each intermediate compartment. The program for the EBN last dendrite compartment, whose descriptive of dynamics is Eq. (2.3), is shown in Fig. 2.11a.

The axon module is the core drive module of the program. The saccade-induced spiking activities at the premotor level are modeled with an HH model for the bursting neurons [Enderle and Zhou, 2010]. The tonic spiking behavior of the TN/IN is implemented by a modified FHN model as well [Faghih et al., 2012]. The program for the EBN axon module is presented in Fig. 2.12a. The underlying differential equation is Eq. (2.4) that conveys the dynamics of the membrane potential. The program for the M differential equation, with the parameters defined in Eq. (2.5), is depicted in Fig. 2.12a. After the simulation is executed at this level, the results of simulation indicate if the burst activity timing and peak rate are as desired for each neuron. Recall that transmission along the axon introduces a time delay, subsequent to which an action potential appears at the synapse.

At the synapse level, each neuron sends out a pulse train stimuli to the postsynaptic neurons in the context of the neural connections in Fig. 2.9. Synaptic connections between functionally modeled neurons are modeled following a current-based synapse scheme [Wong et al., 2012]. Recall that in this scheme, the time course of the model is based on the expressions given in Eqs. (2.7)—(2.9). The EBN synapse module of the program is displayed in Fig. 2.13. The switch block uses a threshold to convert each action potential to a current pulse with the amplitude defined in Table 2.2. The output pulse width can be altered by changing the threshold.

The programs for the agonist and antagonist neural controllers are presented in Fig. 2.14. From the burst activity of the saccade neural network, the agonist neural input is the firing rate of the ipsilateral AN, and the firing rate of the ipsilateral ON is the antagonist neural input. As presented in the previous chapter, the estimation of these two neural inputs was achieved by model predictions. The obtained active-state tensions here from Fig. 2.14 are an alternative for driving the program depicted in Fig. 1.8b. The agonist and antagonist tonic inputs from inactive motoneurons are modeled as illustrated in Fig. 1.8g and h, respectively. Shown in Fig. 2.15 is the program that justifies the relationship between the outputs of the neural controllers—active-state tensions—and the muscle fiber models illustrated in Fig. 1.8a–h. It is worth noting that the motoneuronal firing rates (input blocks in Fig. 2.14) are herein obtained from the simulation of a network of biophysical neurons in the midbrain.

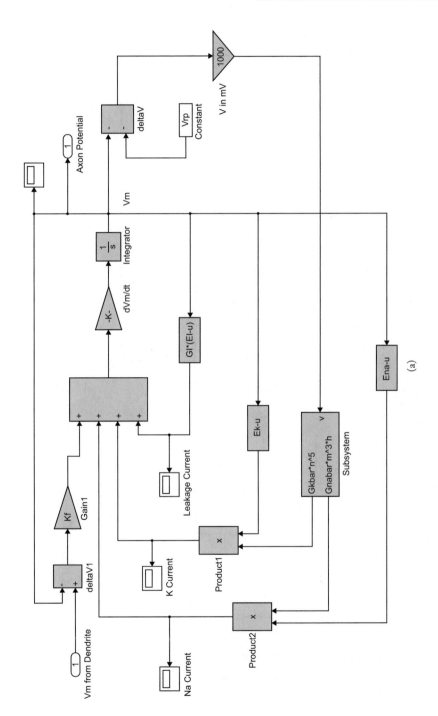

Figure 2.12: The EBN axon module of the program. (a) The main module including the subsystem of N, M, and H coefficients. (*Continues.*)

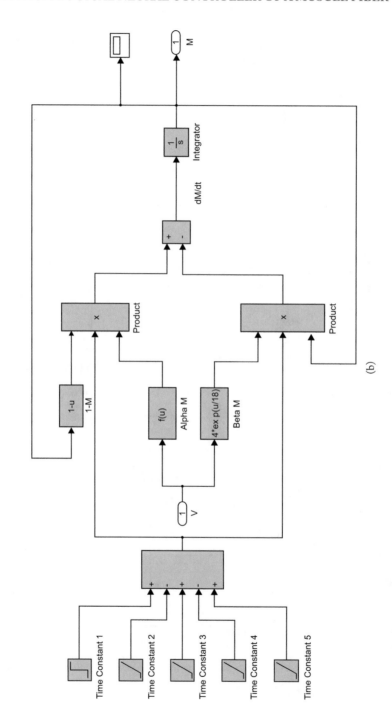

(b)

Figure 2.12: *(Continued.)* The EBN axon module of the program. (b) The *M* differential equation of the sodium current block units.

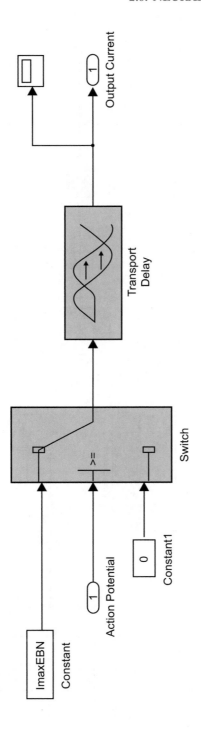

Figure 2.13: The EBN synapse module of the program.

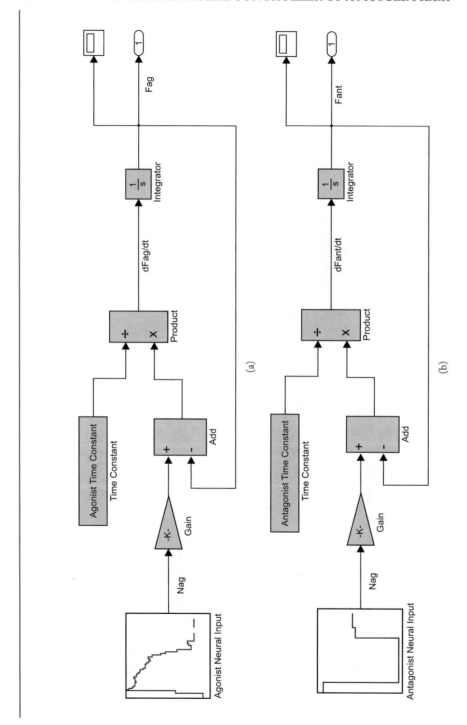

Figure 2.14: The time–optimal neural controller program. (a) The ipsilateral agonist neural contontroller program. (b) The ipsilateral antagonist neural controller program.

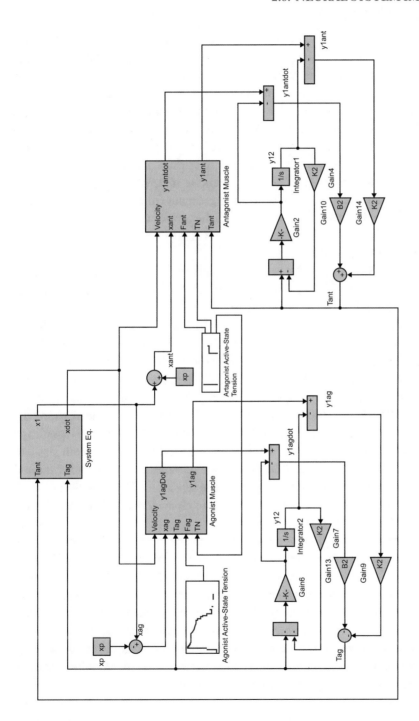

Figure 2.15: Program showing the interaction between the neural controllers and the muscle fiber models. The agonist active–state tension is the input to each agonist muscle fiber, simultaneously to which an antagonist muscle fiber is stimulated by its corresponding active–state tension. The active–state tension modules are inserted from the program depicted in Fig. 2.14. The difference between this program and the one in Fig. 1.8b is that the former implements a network of neurons, rather than model predictions of the motoneuronal inputs, to yield the muscle innervation signals at the motor stage of the neural network.

2.6.2   CONTROL SIMULATION RESULTS

The saccade-induced spiking activities at the premotor level are modeled with an HH model for the bursting neurons [Enderle and Zhou, 2010]. The onset delay before saccade, peak firing rate, and burst termination time for the different neuron populations are chosen according to Table 2.1.

As suggested in the previous chapter, we use 100 identical muscle fibers ($n = 1$ and $m = 100$), since this coordination provided sufficient resolution in matching the experimental data. As described, the number of active neurons impacts the control of saccades instead of the variations in the firing rate of those neurons under the time-optimal control strategy. In addition, the number of active neurons differs from saccade to saccade, as evident by the dynamics observed in the main-sequence diagrams. As demonstrated, this system parameter is determined by reducing from a maximum of 100 active neurons until the eye position estimate from the MFM and the whole muscle model match. The active-state tension for each of the agonist neurons that are not activated is modeled to exponentially decay (during the pulse) and rise (during the slide) using the same time constants in the agonist controller model.

As results proved in Section 1.5, the neural innervations from this number of neurons for each small saccade of the muscle fiber oculomotor plant tend to be in excellent agreement with those of the whole muscle oculomotor plant. Each active neuron exhibits the pause-slide-step firing trajectory, as later shown in Fig. 2.18, substantiating the physiological accuracy of the agonist controller model. The adjustment of the number of active neurons for the large saccades is empirically carried out to maximize the correlation between the whole muscle oculomotor plant and the muscle fiber oculomotor plant in the previous chapter. As such, the number of active agonist neurons for the 4° and 8° saccades is 48 and 76, respectively. This number is also estimated to be 75 neurons for the 12° saccade, 100 neurons for the 16° saccade, and 92 neurons for the 20° saccade. Table 2.3 lists the number of active neurons, and the duration of the burst (agonist pulse), for the five different saccades herein. Notice that the latent period is not zero in our simulations. The saccades start at 120 ms. The termination time of the saccades solely depends on the duration of burst under the time-optimal control strategy. The selection of the duration of the burst is in accord with the saccade duration-saccade magnitude characteristic of the main-sequence diagrams [Enderle and Zhou, 2010].

For sample illustrations, the plots of dendritic membrane potential (first column), axonal membrane potential (second column), and synaptic current pulse train (third column) for the burst neurons and the IN in generation of the 16° saccade are shown in Fig. 2.16. Recall that the train of action potentials is converted to a train of the current pulses in the presynaptic terminal of the neuron to provide excitatory or inhibitory input to the succeeding neurons based on the neural connections in Fig. 2.1. This current pulse flows through the postsynaptic dendritic compartments of the latter neurons, thus providing the smooth postsynaptic potentials to prime the axonal compartment. It is evident that, upon the increasing of the stimulus current pulse amplitude, the depolarization of the postsynaptic membrane intensifies.

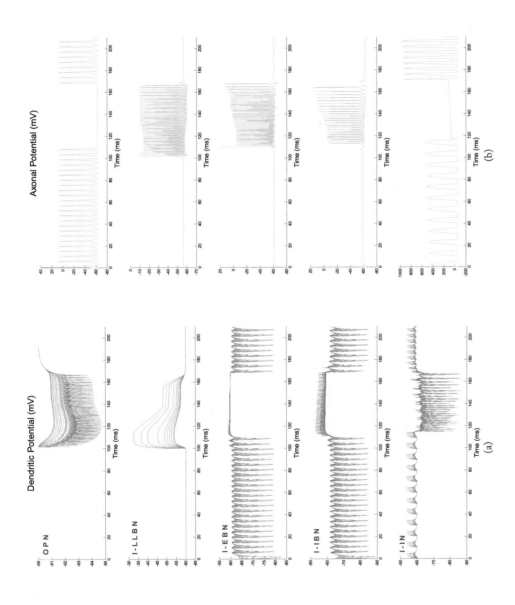

Figure 2.16: The dendritic membrane potential in mV (a), and axonal membrane potential in mV (b). OPN (top), ipsilateral LLBN, EBN, IBN, and IN are shown in order. *(Continues.)*

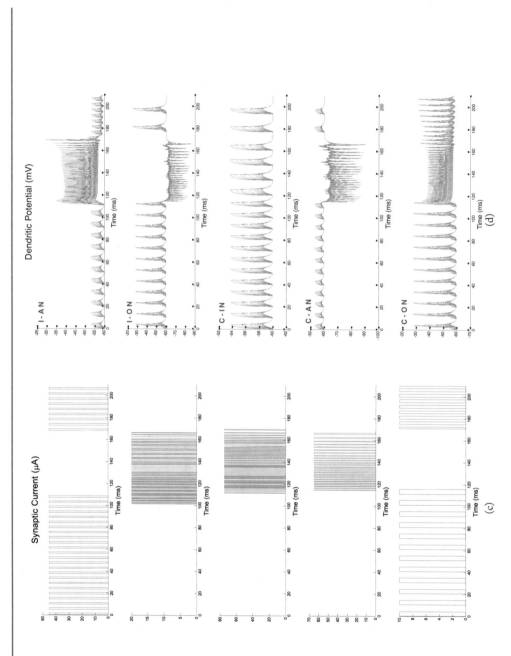

Figure 2.16: *(Continued.)* The synaptic pulse current train in μA (c) of five neurons in a 16° saccade neural controller and the dendritic membrane potential in mV (d). *(Continues.)*

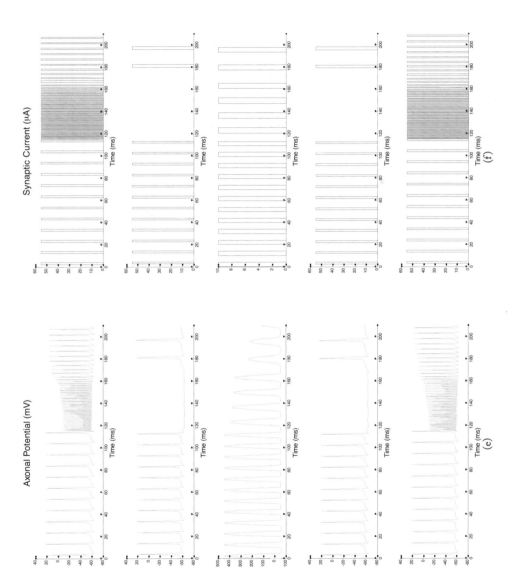

Figure 2.16: *(Continued.)* The axonal membrane potential in mV (e) and the synaptic pulse current train in μA (f) of five neurons in a 16° saccade neural controller. Shown in consecutive rows are ipsilateral AN and ON, as well as contralateral IN, AN, and ON. Each neuron fires in harmony with the others in generating this saccade (saccade onset: 120 ms).

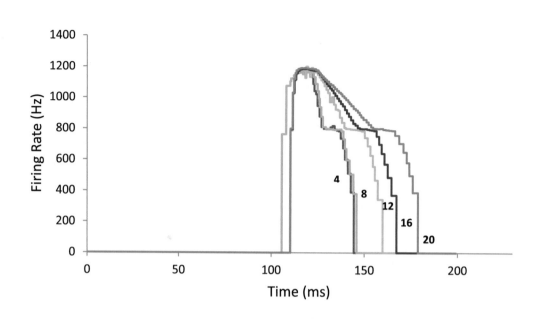

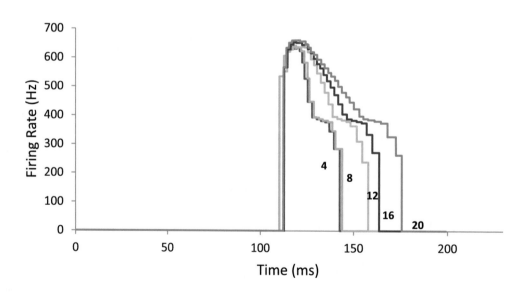

Figure 2.17: Sample burst firing trajectories during the pulse interval of innervation in five saccades. (Top) Ipsilateral EBN. (Bottom) Ipsilateral IBN.

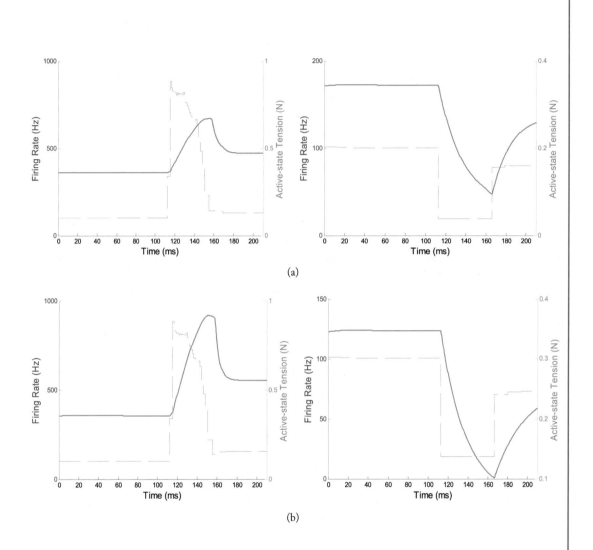

Figure 2.18: The ipsilateral neural stimulation signals for the agonist (first column) and antagonist (second column) neural control inputs (dashed), and the corresponding active-state tensions (solid) plotted on the same graph. (a) 4° saccade and (b) 8° saccade. *(Continues.)*

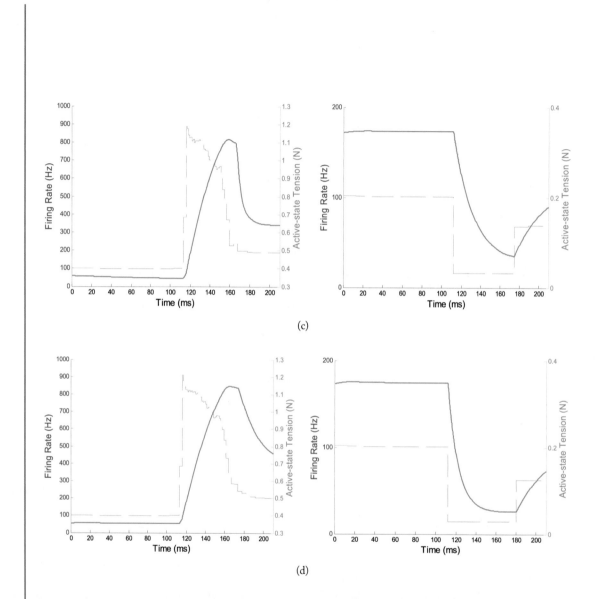

Figure 2.18: *(Continued.)* (c) 12° saccade and (d) 16° saccade. *(Continues.)*

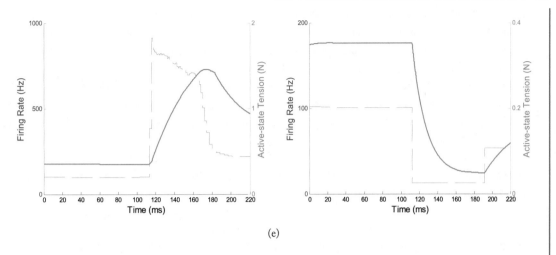

(e)

Figure 2.18: *(Continued.)* (e) 20° saccade.

Table 2.3: Time-optimal control of the saccade magnitude with the duration of burst firing and the number of active neurons

Saccade Magnitude (Degrees)	Agonist Pulse Duration (ms)	Number of Active Neurons
4	40	48
8	42	76
12	52	75
16	56	100
20	65	92

It is obvious that the synapse propagation raises different excitatory and inhibitory postsy-naptic potentials in the dendritic compartments of each postsynaptic neuron (shown in the first column of Fig. 2.16). One can realize that, in view of the trajectory of changes in the membrane potential among the compartments, each postsynaptic neuron, in turn, can either become closer to firing an action potential chain, or inhibited from firing. It is clear that as the presynaptic input pulses are closely spaced in time, each succeeding postsynaptic potential is smaller than the basic single-pulse response, but the postsynaptic response to each input pulse is demonstrable.

It is worthy of note that the ipsilateral LLBN membrane response is different from the rest, since it is stimulated by the contralateral SC current pulse, as shown in Fig. 2.2. Note that the EBN serves as the fundamental excitatory input for the analysis of the saccade controllers. When the ipsilateral EBN is weakly stimulated by the contralateral FN, it renders a special membrane property that tends to a high-frequency burst mechanism until inhibition from the contralateral IBN and the OPN. The burst firing trajectory of the ipsilateral EBN and IBN for saccades of all sizes is presented in Fig. 2.17. It can be seen that these neurons start burst firing at very high levels approximately 8 ms before the saccade starts (see Table 2.1). The onset of the second portion of the burst in all cases is 125 ms. The gradual decay in firing occurs in the interval from this instant until approximately 10 ms before the neurons stop firing. The mechanism for modeling this decay in firing in the axon model is to reduce the firing rate linearly by modifying the channel equations, as mentioned previously. It is noteworthy that the only difference between the three saccades is the duration of the. second portion of the burst, by the end of which the EBN drives the motoneurons to move each eye to its destination.

Presented in Fig. 2.18 are the ipsilateral agonist and antagonist firing rates with their respective active-state tensions based on the agonist and antagonist controller models. Evidently, the burst-tonic firing activity of the agonist neurons reflects the burst firing of the EBN along with the tonic firing of the IN. It is of interest to note that the firing rate of each agonist neuron does not vary as a function of saccade magnitude in any case. This observation shows that the proposed time-optimal controller is fairly capable of mimicking the physiological properties of the saccade by merely changing the duration for the saccades. The agonist and antagonist active-state tensions during the periods of fixation are found as functions of eye position at steady-state [Enderle and Zhou, 2010]. From Fig. 2.18, it also follows that the agonist-antagonist firing patterns fairly well match the estimated waveforms based on the system identification approach, described in Section 1.4. In particular, the firing trajectory of the agonist neural input approximates the burst-tonic data during the pulse and slide intervals of innervation accurately.

The ipsilateral control simulation results of eye position for the two small saccades under the time-optimal control strategy are demonstrated in Fig. 2.19. The position trajectories are all congruent with those achieved by using parameter estimations in Section 1.5. The trend of changes in muscle tensions involved in each saccade is such that neuron-data-derived active-state tensions drive the muscle fiber oculomotor plant.

Shown in Fig. 2.20 are the ipsilateral control simulation results for the three large saccades under the time-optimal control strategy. It is of interest to note that, as envisioned [Enderle and Wolfe, 1987, Zhou et al., 2009], the investigated oculomotor plant does not considerably influence the main-sequence diagrams. Comparison of the obtained saccade characteristics with the analytical solutions in Section 1.5 demonstrates remarkable consistency. It is noted, however, that even for the saccades of the same magnitude, there could be recognizable differences in the latent period, time to peak velocity, peak velocity, and peak acceleration. Hence, it is known that saccades of the same magnitude usually exhibit different trajectories. It proves fundamental, nonetheless, that the time-optimal controller fairly well accommodates this variability. The entire eye movement trajectories (position, velocity, and acceleration) on the contralateral side were in close agreement with their corresponding ipsilateral signals for all of the saccades.

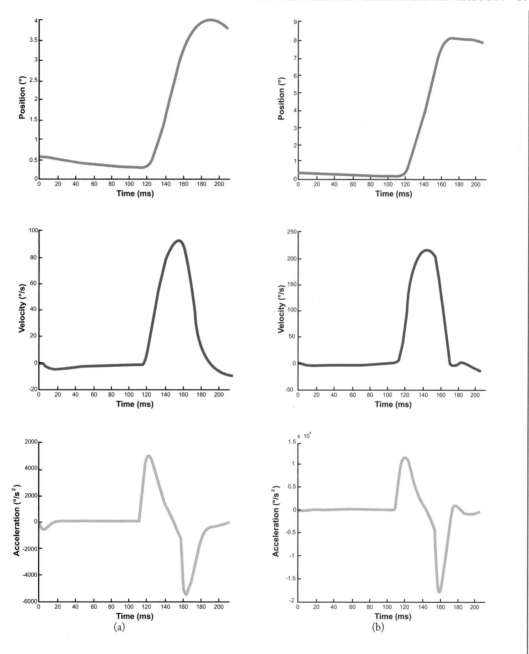

Figure 2.19: The ipsilateral control simulation results (position, velocity, and acceleration) for the monkey small saccades generated by the time-optimal neural controller in the muscle fiber oculomotor plant. (a) 4° saccade. (b) 8° saccade. Note that the saccade onset is 120 ms for all cases, but the end time of each saccade differs from the rest.

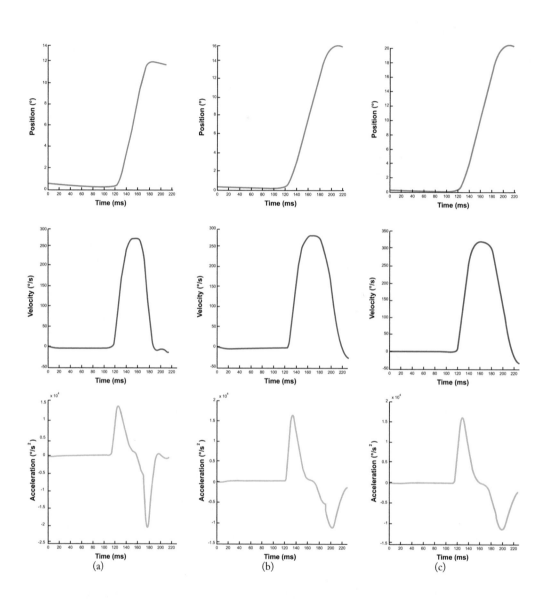

Figure 2.20: The ipsilateral control simulation results (position, velocity, and acceleration) for the monkey large saccades generated by the time-optimal neural controller in the muscle fiber oculomotor plant. (a) 12° saccade. (b) 16° saccade. (c) 20° saccade. Note that the saccade onset is 120 ms for all cases, but the end time of each saccade differs from the rest.

2.7   DISCUSSION

The simulation results show remarkable agreement with those provided by analytical descriptions of the agonist and antagonist neural inputs, and the corresponding active-state tensions for the small saccades (see Section 1.5). The trajectory of variation in the agonist pulse magnitude among these saccades is consistent with the agonist pulse magnitude-saccade magnitude characteristic for the large saccades [Enderle and Zhou, 2010]. The burst duration is found to show similar correlation to the MLBN duration of burst firing from the extracellular single-unit recordings [Sparks et al., 1976].

2.7.1   NEURAL NETWORK FRAMEWORK

As evident by different firing rate trajectories for the EBN, this neuron has characteristics that are tightly coupled to the saccade. For the three saccades examined herein, the initial duration of the EBN firing remained constant among them. However, the duration of the second portion of the burst discharge (gradual drop) varied among them, based on the entire duration of the burst firing in Table 2.3. As indicated in Table 2.1, the EBN firing lags behind the saccade by 6–8 ms, whereas the AN starts burst firing 5 ms before the saccade. Finding the dendrite parameters for both of these neurons in meeting the required onset time delay was tedious. Moreover, the AN peak firing rate at the beginning of the pulse period showed dependency on the EBN peak firing rate, necessitating the use of corresponding coefficients in Table 2.2 to change the initial firing rate of the basic HH model.

Implementing the OPN dendrite and synapse models in order that this neuron stops inhibiting the EBN about 10 ms before the saccade, and resumes its inhibition almost when the saccade ends, was subject to numerous parameter tunings. Without this coordination in timing of the burst firing in the EBN, this neuron can show the rebound burst firing activity. This rebound burst, in turn, causes the saccade to deviate from the normal characteristics. It was also vital that the end of the IBN inhibition of the antagonist motoneurons coincides with the resumption of tonic firing in them such that no deviation from the normal saccade, or truncation of saccade, is present.

Zhai et al. coordinated experiments with goal-oriented saccades to trigger human saccades—from a pool of subjects—using a high-speed eye tracking system [2013b]. Types of targets included visual, auditory, and auditory-visual bisensory stimuli. The ensuing saccade characteristics were analyzed and compared in depth. The experimental findings set forth that: (1) the auditory-visual stimuli provided the highest saccade accuracy; saccade peak velocity increased up to 700 °/s in an exponential manner, as saccade amplitude accrued; (2) saccade duration was approximately constant for small saccades under 7°, whereas it was linearly proportional to saccade amplitude for large saccades—among the responses, auditory saccades showed lower peak velocity and longer duration; (3) saccade latent period was around 100–300 ms and was relatively independent of saccade amplitude, but a significant reduction of the latent period was observed in the bisensory cases; and (4) there was a higher probability of occurrence of dynamic overshoot in auditory saccades; in particular, more in the abducting direction than the adducting direction.

Coubard [2013] reviewed different lines of controversy in research between the proponents of binocular coordination of eyes vs. those of monocular coordination during combined saccade-vergence eye movements. It is suggested that, in order to fully respond to target displacements in all feasible depths and directions, saccade-vergence commands can be simultaneously processed by both eyes, as well as received individually in monocular fashion, especially in view of the neurophysiological manifestations. The treatment to modeling the pure saccades on the basis of the local feedback model [Zee et al., 1992] has been the focus of attention. In this model, a conjugate saccadic velocity command, derived from the saccade burst neuron model, is modulated through local filters to provide the oculomotor motoneurons with the pulse, slide, and step of innervation. In the generation of an ipsilateral saccade, the pulse force of the premotor neurons is attributed to the burst discharge within the PPRF, whereas their step force is related to the burst discharge in the bilateral nucleus prepositus hypoglossi and in the medial vestibular nucleus [Scudder et al., 2002]. The OPNs tonically inhibit the premotor neurons as early as the saccade terminates. The premotor commands finally flow through the ANs that innervate the ipsilateral lateral rectus muscle, when the same innervation is exerted to the contralateral medial rectus muscle by intervention from the abducens internuclear neurons.

While the midbrain coordination mechanism in generating saccades has been qualitatively studied [Walton et al., 2005] and [Coubard, 2013, Enderle, 1994, Girard and Berthoz, 2005], a complete neural circuitry that includes both the premotor and motor neurons in characterizing the final motoneuronal command to the extraocular muscles has not yet been developed. The utility of SNNs to the biophysical modeling of interconnected neurons [Ghosh-Dastidar and Adeli, 2007, 2009, Mohemmed et al., 2012] elucidates broad principles to modeling at higher structural scales, such as the circuit [Enderle and Zhou, 2010, Rosselló et al., 2009] and the systems levels [Ramanathan et al., 2012]. In this work, a neural circuit model was demonstrated and parameterized to match the firing characteristics of eight neuron populations at both the premotor and motor stages of the neural network. Despite the complexity of the saccade generator in a large-scale SNN, the neural modeling approach led us to address the challenges involved in the implementation of the midbrain pathways due to the ensuing heavy storage and computational requirements.

2.7.2   NEURAL CONTROLLERS

There is some divergence of opinion among researchers as to the extent of dependency of the saccade magnitude on the firing rate in agonist neurons. For instance, a velocity-based controller did not provide satisfactory evidence to permit any conclusion in regard to the agonist firing trajectory observed in the experimental data [Sylvestre and Cullen, 1999]. Investigation of the physiological evidence, however, provides the fact that the agonist neurons' accumulative firing rate peaks at a maximum level, gradually drops to another firing rate level, and in the end levels off at a tonic firing rate [Enderle and Sierra, 2013]. Exactly how there is a one-to-one relationship between the firing rate in agonist neurons and the saccade magnitude is another matter of controversy among the researchers. For example, some aimed to establish firing-rate, saccade-amplitude-dependent controllers [Gancarz and Grossberg, 1998, Scudder, 1988]. These studies lacked the use of a homeomorphic oculomotor plant and did not develop a network of neurons to execute saccades.

A time-optimal neuronal control strategy for saccadic eye movements was first proposed based on experimental data analysis [Clark and Stark, 1975]. In their curve-fitting investigation to match a nonlinear model to saccadic eye movement data, with activation and deactivation time constants, they concluded that the best match is obtained with a first-order pulse-step neuronal controller. Without the use of fitting parameters in their system identification approach, however, they reported a second-order time-optimal controller using optimal control theory. They further proceeded to show that for a reduced fourth-order model, eliminating the activation and deactivation time constants, the optimal control investigation achieved a first-order time-optimal controller. This observation is consistent with their experimental findings. Later on, the neuronal control strategy for human saccadic eye movements was investigated using optimal control theory based on the minimum principle of Pontryagin in a linear model [Lehman and Stark, 1979]. This model included the activation and deactivation time constants, but did not lead to robust results. After eliminating the activation and deactivation time constants, thus lowering the order of the model from sixth- to fourth-order, their analysis resulted in a second-order time-optimal controller. Their simulation of saccades was, however, contrary to the minimum principle of Pontryagin, since they presumably expressed the agonist pulse magnitude as a function of saccade magnitude. As a result, their second-order controller was not found to be time-optimal. It is noteworthy that the exclusion of the activation and deactivation time constants in the optimal control theory approach to derivation of the controllers contradicts with the ample physiological evidence substantiating the inclusion of these time constants in the models [Robinson, 1981, Zhou et al., 2009]. The inclusion of the time constants therefore appears to be substantial for the quantitative analysis of the models of saccades, with little or no sacrifice in optimality. To conclude, none of the investigated controllers offers the feasibility and validity of a time-optimal control strategy analytical solution.

Korentis [2009] presented specifications for design and development of a robotic ophthalmotrope (Robophthalmotrope). The main objective of devising the Robophthalmotrope has been to construct a physiologically consistent, biomimetic, mechatronic platform to examine the theoretical concepts related to the oculomotor system. In the context of a three-dimensional eye movement system, the neural innervations command three pairs of agonist and antagonist muscles to orient the eyeball horizontally, vertically and torsionally in a precise fashion. As the groundwork for its actuator design, Robophthalmotrope, unlike its predecessors, is intended to maintain the agonist and antagonist control actuations separately for each pair of extraocular muscles. In terms of the biomechatronics, it is indicated that a master digital controller will coordinate the signaling of 12 slave microcontrollers (one microcontroller for each actuator). This implementation is envisioned to be effective in enhancing the human-like performance measures in the application of service and sociable robotics.

In this work, a first-order time-optimal controller is used, which includes the activation and deactivation time constants in agonist and antagonist controller inputs to the muscle fiber oculomotor plant. This controller has been proven to agree with the experimental findings [Clark and Stark, 1975, Enderle and Wolfe, 1988]. Realization of the suitable time constants for both the agonist and antagonist controllers was key in providing the required steady-state active-state tensions to the muscle fiber oculomotor plant. The estimated activation and deactivation time constants from the system identification approach by Zhou et al. [2009] best satisfy this specification. Without such appropriate parameters, the simulated saccade could be showing deviations from the intended position at steady-state.

2.7.3   MUSCLE FIBER MODEL AND SACCADES

The set of agonist-antagonist control inputs to the muscle fiber oculomotor plant supports the time-optimal controller in which the motoneurons' firing rate does not determine the saccade magnitude. The application of the MFM in the oculomotor plant proves important in accommodating the constraint on the number of active neurons firing maximally in controlling the saccade magnitude. The number of the active neurons is a key parameter whose adjustment in the MFM is vital in providing the desired saccade control simulation results. Our observations ascertain that the duration of the agonist burst firing and the number of active agonist neurons are integral to determining the saccade size (see Table 2.3).

It is noteworthy that the duration of agonist burst discharge is of prime significance in determining the saccade magnitude as seen in Fig. 2.18. It is concluded that the neural network is constrained by a minimum duration of the agonist pulse, and that the most dominant factor in determination of the amplitude is the number of active neurons for the small saccades. For the large saccades, however, the duration of agonist burst firing is directly related to the saccade magnitude. The number of active neurons for the 16° and 20° saccades remains relatively the same, although the 12° saccade aggregates fewer active neurons as seen in Table 2.3. The discussion in Section 1.6 is enlightening as to the increasing movement field of activity within the SC for saccades up to 12° for the monkey data. Furthermore, from the velocity profiles for the simulated saccades, it was found that monkey saccade has a larger peak velocity than human [Enderle and Sierra, 2013].

The final eye position results establish evidence for the acceptable performance of the proposed neural circuitry and the exploited time-optimal controller in modeling the horizontal monkey saccades. The dependence of these different saccades on the agonist pulse duration has been found to be well presented by our time-optimal controller. The simulation results substantiate the time-optimal controller by the close agreement obtained with the analytical solutions of saccade characteristics [Enderle and Zhou, 2010, Enderle and Sierra, 2013]. This agreement gives rise to the accuracy of the membrane parameters in neural modeling, as listed in Table 2.2.

2.8   DEMONSTRATION OF THE CONJUGATE GOAL-DIRECTED HORIZONTAL HUMAN AND MONKEY SACCADES PROGRAMMING INTERFACE

The implemented Simulink program of the neural system was explained in Section 2.6. A recent study demonstrated the results of the conjugate goal-directed horizontal human saccade [Ghahari and Enderle, 2014]. Furthermore, a major focus of the research has been to describe mathematical model predictions, revealing that the contralateral EBN's post-inhibitory rebound burst activity after marked hyperpolarization causes dynamic overshoots or glissades in humans [Enderle, 2002, Enderle and Zhou, 2010, Zhou et al., 2009]. The unplanned PIRB toward the end of the saccade was well represented by their model. A graphical user interface (GUI), which integrates and invokes the Simulink programs of the human normal saccades and glissades, and monkey saccades, is designed and presented here. The intent is to provide an interface construct for the user to define and test her specified neural system (within a set of parameters). One advantage that this GUI offers is that the user need not be competent with the Simulink software at any simulation event. This GUI can be run on MATLAB R2012a or higher.

2.8.1   GUI DEVELOPMENT FOR SIMULINK PROGRAMS

The Conjugate Goal-Directed Horizontal Human and Monkey Saccades (CGDHHMS) is developed using GUIDE-generated callbacks. It is composed of the components, each of which has a callback property that is associated with the component handles. The GUI callback functions use the 'current' workspace to read/write the data attributes. However, in order to properly execute the Simulink programs within the CGDHHMS, one should set the workspace to the "base." This setting ensures that the Simulink models' data attributes are constructed after each run, and that the invoked ones are passed over from the "base" workspace to the "current" workspace. The handshaking between the GUI components and the Simulink programs, from the simulation start to the GUI update, is achieved by assigning callback functions to the GUI components in invoking the required Simulink programs.

The CGDHHMS hierarchical structure includes a list box, a pop-up menu, and three radio buttons at the input level, and four panels at the demonstration level, in conjunction with two high-end push buttons. The user interacts with the CGDHHMS to design a simulation scenario and view the generated results. The interactions allow the user to abstract the physiological and computational structures of the neural network, thus constructing custom neural models that are capable of generating the final motoneuronal signals to oculomotor plants. To serve this goal, the top-level system specifications can be chosen from the list box and the pop-up menu components, and a selected subset of simulation parameters can be specified from the radio buttons.

The list box has three options for the user to select from which the subject whose Simulink model is to be executed. It is populated by the list of Human Normal, Human Glissade, and Monkey. Once the subject is chosen, the user should specify the magnitude of the desired saccade from the pop-up menu. The string property of the list box is retrieved by the Run push button's callback function.

The pop-up menu provides the user with the different magnitudes option for the simulation. When the user selects an item, the pop-up menu's properties are updated to retrieve the required function callback. The contents of the pop-up menu are magnitudes of 4°, 8°, 10°, 12°, 15°, 16°, and 20°. Among this list are the human saccades (10°, 15°, and 20°), as well as the monkey saccades (4°, 8°, 12°, 16°, and 20°). When the user chooses a magnitude, this value is stored to be retrieved by the Run push button's callback function when executed.

There are three radio buttons to allow the user to specify a subset of parameters of the neural model and the oculomotor plant. First, the dendrite morphological toggle states are 3 compartments and 14 compartments. Second, the options for axon model mechanism are to choose from the Hodgkin Huxley (HH) model and the FitzHugh-Nagumo (FHN) model. Third, the options for oculomotor plant are to choose either the lumped parameter model for the Human or the muscle fiber model for the Monkey. Each radio button's callback function fetches and executes the Simulink model associated with the user-specified system and simulation parameters. CGDHHMS has the capability to parse and restore the simulation parameters in consecutive runs.

Results are demonstrated in four GUI panels. Two of the four panels contain multiple axes. The axes embedded in these two panels showcase the results associated with the neural network simulation of burst neurons and the high-end saccade characteristics. The other two panels contain the first-order time-optimal neural controller results. The handle of each panel is used for executing or updating the plotting commands within the Run push button's callback.

In the context of the above description, Fig. 2.21 shows a block diagram representation of the hierarchical structure of CGDHHMS during user interface. After specifying the system and simulation parameters in user's entries, the interaction between various levels of CGDHHMS starts with generation of the neural commands within the corresponding Simulink program. The ensuing muscle innervation signals are then transferred to one of the muscle models, which, in turn, provide the corresponding saccade responses. The demonstration and interpretation of such responses comprise the lowest level of the GUI development.

2.8.2   CGDHHMS INTERFACE

For a single-run interface, CGDHHMS user can proceed as follows.

1. Select the Subject from the list box at the upper left corner.

2. Specify the Magnitude from the pop-up menu below the list box.

3. Adjust, if necessary, the simulation parameters, each of which are provided in the radio buttons.

4. Push the Run button; wait until the figure panels are populated by the results.

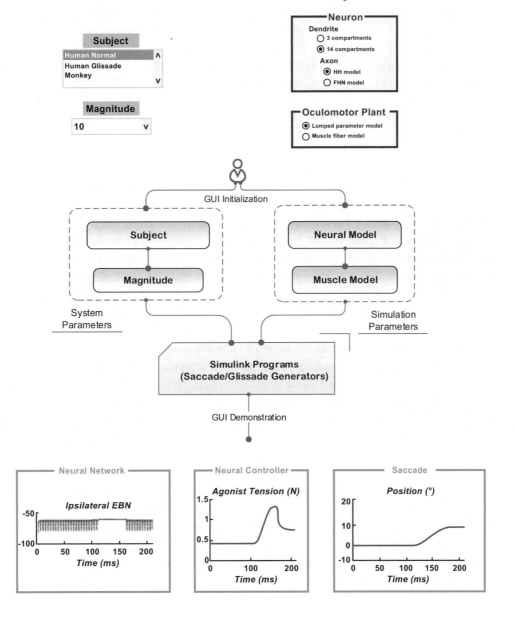

Figure 2.21: User interface with CGDHHMS and its hierarchical layout, from GUI initialization to the demonstration of the results. The user first specifies the system parameters: Subject and Magnitude. The simulation parameters, regarding the neural model and the oculomotor plant, are chosen next. After this specification of the parameters is complete, the corresponding Simulink programs are called and executed. Consequently, different figure panels will exhibit some characteristics of the user-specified neural system.

After each display of the results has been viewed, the demonstrated data can be saved by issuing the Save As command under the File menu. There are also other functionalities placed in the GUIDE Menu Editor for purpose of enhancing viewing and saving of the results. Sample clipboard images of various user's trials with CGDHHMS are illustrated in Fig. 2.22. We will continue to further enhance the usability and add to the scalability of CGDHHMS in the future stages of its development.

2.9 CONCLUSION

We simulated five different conjugate goal-directed horizontal monkey saccades: 4°, 8°, 12°, 16°, and 20°. A parallel-distributed neural network model of the midbrain was first presented. To develop the quantitative computational models that establish the basis of this functional neural network model, the saccade burst generator dynamics were described next.

This work investigated an integrative systems approach to address the challenges involved in the implementation of the saccade dynamics from the local neural circuits due to the ensuing heavy storage and computational requirements. Neural circuitry, including omnipause neuron, premotor excitatory and inhibitory burst neurons, long lead burst neuron, tonic neuron, interneuron, abducens nucleus, and oculomotor nucleus, was developed to examine such saccade dynamics. An optimal control mechanism demonstrated how the neural commands were encoded in the downstream saccadic pathway between superior colliculus and motoneurons. Ultimately, the horizontal monkey saccades were well characterized by integrating the neural controllers to a linear homeomorphic muscle fiber oculomotor plant. 100 identical muscle fibers were connected in series in both the agonist and antagonist muscles in the oculomotor plant. Under the time-optimal strategy, the number of neurons that actively fire and the duration of the agonist pulse determined the saccade magnitude. The choice of the number of active neurons proved accurate in adapting the muscle fiber model to provide the desired control simulation results.

A physiologically based model of the neuron was modeled herein, for which a program simulated the underlying membrane differential equations. The ensuing computational cost was reasonable because of the rationalized modular programming of the neural network. The proposed saccadic system thus presented a complete model of saccade generation, since it not only included the neural circuits at both the premotor and motor stages of the saccade generator, but it also used a time-optimal controller to yield the desired saccade magnitude.

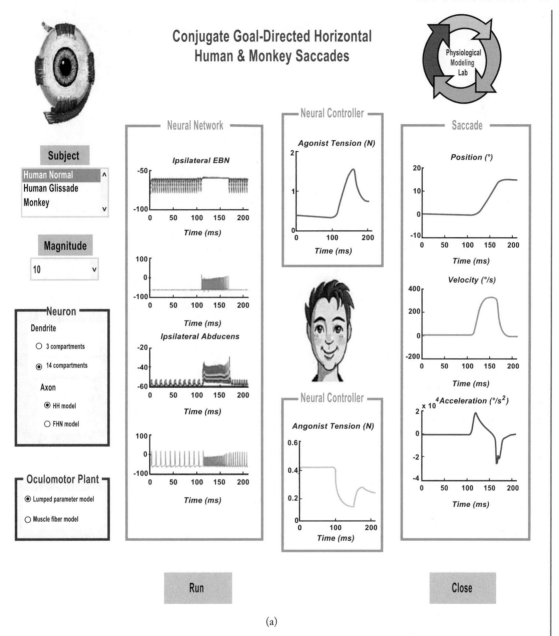

(a)

Figure 2.22: Simulation results provided for a single run of CGDHHMS. User-specified parameters are shown on the left pane. The results are illustrating some membrane potential trajectories from the neural network, active-state tensions from the neural controller, and the saccade characteristics. (a) For a 10° Human Normal saccade. *(Continues.)*

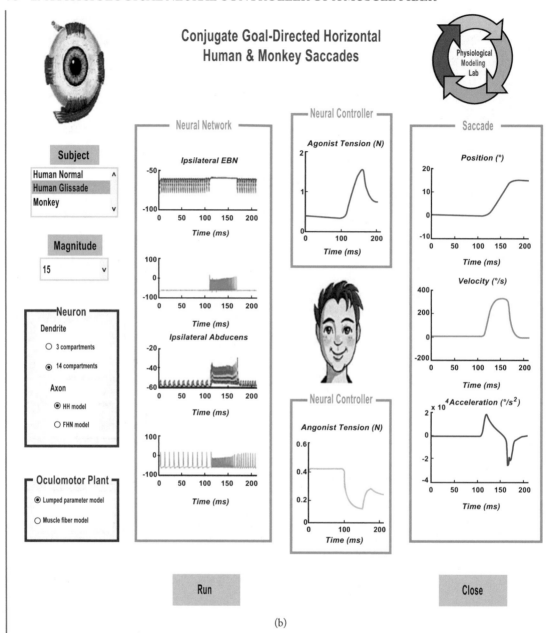

(b)

Figure 2.22: *(Continued.)* Simulation results provided for a single run of CGDHHMS. User-specified parameters are shown on the left pane. The results are illustrating some membrane potential trajectories from the neural network, active-state tensions from the neural controller, and the saccade characteristics. (b) For a 15° Human Glissade. *(Continues.)*

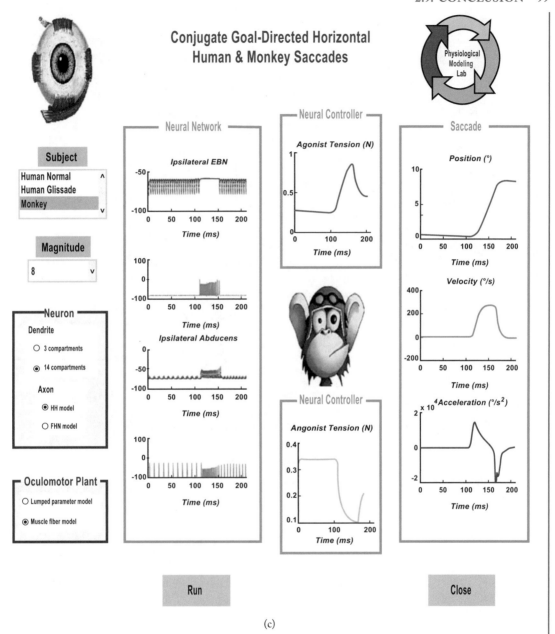

(c)

Figure 2.22: *(Continued.)* Simulation results provided for a single run of CGDHHMS. User-specified parameters are shown on the left pane. The results are illustrating some membrane potential trajectories from the neural network, active-state tensions from the neural controller, and the saccade characteristics. (c) For an 8° Monkey saccade. *(Continues.)*

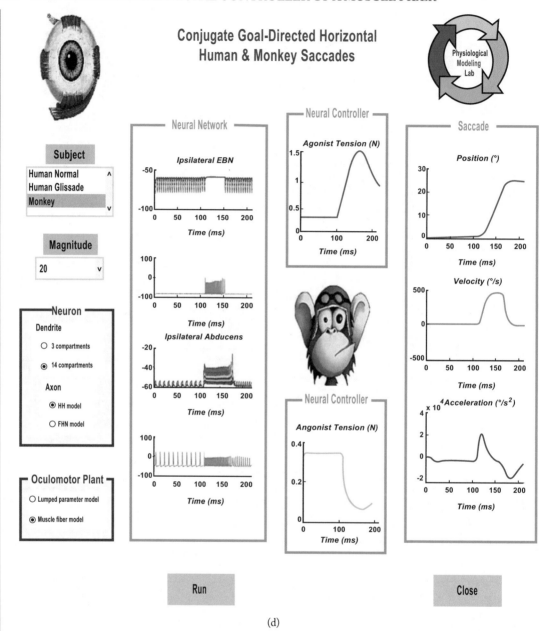

(d)

Figure 2.22: *(Continued.)* Simulation results provided for a single run of CGDHHMS. User-specified parameters are shown on the left pane. The results are illustrating some membrane potential trajectories from the neural network, active-state tensions from the neural controller, and the saccade characteristics. (d) For a 20° Monkey saccade.

Bibliography

Carpenter, R.H.S. (1988) *Movements of the Eyes*, 2nd ed., Pion, London. 3

Clark, M.R. and Stark, L. (1975) Time optimal control for human saccadic eye movement. *IEEE Trans. Automat. Contr.*, AC-20: 345–348. DOI: 10.1109/TAC.1975.1100955. 91, 92

Coubard, O.A. (2013) Saccade and vergence eye movements: a review of motor and premotor commands. *Eur. J. Neurosci.*, 38: 3384–3397. DOI: 10.1111/ejn.12356. 42, 90

Cullen, K.E., Rey, C.G., Guitton, D., and Galiana, H.L. (1996) The use of system identification techniques in the analysis of oculomotor burst neuron spike train dynamics. *J. Comput. Neurosci.*, 3 (4), 347–368. DOI: 10.1007/BF00161093. 32

Enderle, J.D. and Wolfe, J.W. (1987) Time-optimal control of saccadic eye movements. *IEEE Trans. on Biomed. Eng.*, BME-34(1), 43–55. DOI: 10.1109/TBME.1987.326014. 46, 62, 86

Enderle, J.D. and Wolfe, J.W. (1988) Frequency response analysis of human saccadic eye movements: estimation of stochastic muscle forces. *Comp. Bio. Med.*, 18: 195–219. DOI: 10.1016/0010-4825(88)90046-7. 62, 92

Enderle, J.D., Engelken, E.J., and Stiles, R.N. (1991) A comparison of static and dynamic characteristics between rectus eye muscle and linear muscle model predictions. *IEEE Trans. Biomed. Eng.*, 38:1235–1245. DOI: 10.1109/10.137289. 2, 7, 8, 10, 36, 41

Enderle, J.D. (1994) A physiological neural network for saccadic eye movement control. Air Force material command. *Armstrong Laboratory AL/AO-TR-1994–0023*: 48 pages. DOI: 10.1155/2014/406210. 42, 46, 90

Enderle, J.D. and Engelken, E.J. (1995) Simulation of oculomotor post-inhibitory rebound burst firing using a Hodgkin-Huxley model of a neuron. *Biomed. Sci. Instrum.*, 31: 53–58. 42, 46

Enderle, J.D. (2002) Neural control of saccades. In J. Hyönä, D. Munoz, W. Heide and R. Radach (Eds.), *The Brain's Eyes: Neurobiological and Clinical Aspects to Oculomotor.* 16, 32, 40, 42, 46, 93

Enderle, J.D. and Zhou, W. (2010) *Models of Horizontal Eye Movements. Part 2: A 3rd - Order Linear Saccade Model*, Morgan & Claypool Publishers, San Rafael, CA, 144 pages. DOI: 10.2200/S00264ED1V01Y201003BME035. 1, 11, 13, 14, 16, 19, 20, 27, 32, 33, 36, 40, 41, 42, 43, 44, 45, 46, 48, 49, 50, 51, 52, 53, 54, 57, 59, 60, 63, 64, 72, 78, 86, 89, 90, 92, 93

Enderle, J.D. and Bronzino, J.D. (2011) *Introduction to Biomedical Engineering 3rd ed.* Elsevier, Amsterdam. 46, 53, 54

Enderle, J.D. and Sierra, D.A. (2013) A new linear muscle fiber model for neural control of saccades. *Int. J. of Neural Systems*, 23 (2), DOI: 10.1142/S0129065713500020. DOI: 10.1142/S0129065713500020. 46, 91, 92

Faghih, R.T., Savla, K., Dahleh, M.A., and Brown, E.N. (2012) Broad range of neural dynamics from a time-varying FitzHugh–Nagumo model and its spiking threshold estimation. *IEEE Trans. on Biomed. Eng.*, 59 (3): 816–823. DOI: 10.1109/TBME.2011.2180020. 41, 46, 54, 72

Gancarz, G. and Grossberg, S. (1998) A neural model of the saccade generator in reticular formation. *Neural Netw.*, vol. 11: pp. 1159–1174. DOI: 10.1016/S0893-6080(98)00096-3. 32, 48, 91

Ganz, J.C. (1988) *Research*, Progress in Brain Research, V. 140, Elsevier, Amsterdam, 21–50.

Ghahari, A. and Enderle, J.D. (2014) A neuron-based time-optimal controller of horizontal saccadic eye movements. *Int. J. of Neural Systems*, Vol. 24, 1450017 (19 pages), DOI: 10.1142/S0129065714500178. 93

Ghahari, A. and Enderle, J.D. (2014) A physiological neural controller of a muscle fiber oculomotor plant in horizontal monkey saccades. *ISRN Ophthalmology*, Article ID 406210. DOI: 10.1155/2014/406210.

Ghosh-Dastidar, S. and Adeli, H. (2007) Improved spiking neural networks for EEG classification and epilepsy and seizure detection. *Integr. Comput.-Aided Eng.*, 14: 187–212. 40, 53, 90

Ghosh-Dastidar, S. and Adeli, H. (2009) Spiking neural networks. *Int. J. of Neural Systems*, 19 (4), 295–308. DOI: 10.1142/S0129065709002002. 40, 53, 90

Girard, B. and Berthoz, A. (2005) From brainstem to cortex: Computational models of saccade generation circuitry. *http://www.ncbi.nlm.nih.gov/pubmed/16343730*, 77(4), 215–51. DOI: 10.1016/j.pneurobio.2005.11.001. 42, 90

Goldberg, S.J., Wilson, K.E., and Shall, M.S. (1997) Summation of extraocular motor unit tensions in the lateral rectus muscle of the cat. *Muscle and Nerve*, 20:1229–1235. DOI: 10.1002/(SICI)1097-4598(199710)20:10%3C1229::AID-MUS4%3E3.0.CO;2-E. 3

Goldstein, H. (1983) The neural encoding of saccades in the rhesus monkey (Ph.D. dissertation). Baltimore, MD: The Johns Hopkins University. 13

Harmon, L.D. (1961) Studies with artificial neurons, I: Properties and functions of an artificial neuron. *Kybernetik* Heft 3(Dez.): 89–101. DOI: 10.1007/BF00290179.

Harwood, M.R., Mezey, L.E., and Harris, C.M. (1999) The spectral main sequence of human saccades. *J. Neurosci.*, 19(20): 9098–9106. 62

Hodgkin, A.L., Huxley, A.F., and Katz, B. (1952a) Measurement of current-voltage relations in the membrane of the giant axon of *Loligo. J. Physiol.*, 116: 424–448.

Hodgkin, A.L. and Huxley, A.F. (1952b) A quantitative description of membrane current and its application to conduction and excitation in nerve. *J. Physiol.* 117:500. 40

Hu, X., Jiang, H., Gu, C., Li, C., and Sparks, D. (2007) Reliability of oculomotor command signals carried by individual neurons. *PNAS*, 8137–8142. DOI: 10.1073/pnas.0702799104. 33, 48

Izhikevich, E.M. (2003) Simple model of spiking neurons. *IEEE Trans. Neural Netw.*, 14: 1569–1572. DOI: 10.1109/TNN.2003.820440. 40

Keller, E.L., McPeek,R.M., and Salz, T. (2000) Evidence against direct connections to PPRF EBNs from SC in the monkey. *J. Neurophys.*, 84(3): 1303–13. 48

Korentis, A. (2009) Robophthalmotrope: Proposing a biologically inspired, mechatronic platform to study sensorimotor control, *Proceedings of the Technologies for Practical Robot Applications*, Woburn, MA, pp. 230–235, http://ieeexplore.ieee.org/xpl/mostRecentIssue.jsp?punumber=5335443 http://dx.doi.org/10.1109/TEPRA.2009.5339617DOI: 10.1109/TEPRA.2009.5339617. 91

Lehman, S. and Stark, L. (1979) Simulation of linear and nonlinear eye movement models: sensitivity analyses and enumeration studies of time optimal control. *J. Cybernet. Inf. Sci.* (2) 21–43. 91

Leigh, R.J. and Zee, D.S. (1999) *The Neurology of Eye Movements*. Oxford University Press Inc., New York, New York. 3, 10

Mohemmed, A., Schliebs, S., Matsuda, S., and Kasabov N., (2012) SPAN: Spike Pattern Association Neuron for learning spatio-temporal spike patterns. *Int. J. of Neural Systems*, 22(4): 1–16. DOI: 10.1142/S0129065712500128. 40, 53, 90

Moschovakis, A.K., Scudder, C.A., and Highstein, S.M. (1996) The microscopic anatomy and physiology of the mammalian saccadic system. *Prog. Neurobiol.*, Oct 50(2–3): 133–254. DOI: 10.1016/S0301-0082(96)00034-2. 48

Olivier, E., Grantyn, A., Chat, M., and Berthoz, A. (1993) The control of slow orienting eye movements by tectoreticulospinal neurons in the catbehavior, discharge patterns and underlying connections. *Exp. Brain Res.*, 93: 435–449. DOI: 10.1007/BF00229359. 48

Optican, L.M. and Miles, F.A. (1985) Visually induced adaptive changes in primate saccadic oculomotor control signals. *J. Neurophysiol.*, 54(4), 940–958. 13

Ramanathan, K., Ning, N., Dhanasekar, D., Guoqi, L., Luping, S., and Vadakkepat, P. (2012) Presynaptic learning and memory with a persistent firing neuron and a habituating synapse: A model of short term persistent habituation. *Int. J. of Neural Systems*, 22(4): 1250015. DOI: 10.1142/S0129065712500153. 40, 53, 90

Raybourn, M.S. and Keller, E.L. (1977) Colliculoreticular organization in primate oculomotor system. *J. Neurophysiol.*, 40: 861–878. DOI: 10.1142/S0129065712500153. 48

Robinson, D.A. (1981) Models of mechanics of eye movements. In: B.L. Zuber (Ed.). *Models of Oculomotor Behavior and Control* (pp. 21-41). Boca Raton, FL: CRC Press. 16, 20, 27, 91

Rosselló, J.L., Canals, V., Morro, A., and Verd, J. (2009) Chaos-based mixed signal implementation of spiking neurons. *Int. J. of Neural Systems*, 19(6): 465–471. DOI: 10.1142/S0129065709002166. 40, 53, 90

Roy, G. (1972) A simple electronic analog of the squid axon membrane. *IEEE Trans. Biomed. Eng.*; 19(1): 60–3. DOI: 10.1109/TBME.1972.324161. 57

Scudder, C.A. (1988) A new local feedback model of the saccadic burst generator. *J. Neurophysiol.*, 59(5): 1455–1475. 32, 91

Scudder, C.A., Kaneko, C., and Fuchs, A. (2002) The brainstem burst generator for saccadic eye movements: a modern synthesis. *Exp. Brain Res.*, 142: 439–462. DOI: 10.1007/s00221-001-0912-9. 90

Sparks, D.L., Holland, R., and Guthrie, B.L. (1976) Size and distribution of movement fields in the monkey superior colliculus, *Brain Res.*, 113: 21–34. DOI: 10.1016/0006-8993(76)90003-2. 33, 34, 89

Sparks, D.L. (2002) The brainstem control of saccadic eye movements. *Neuroscience*, 3: 952–964. DOI: 10.1038/nrn986. 4, 5

Stanton, G.B., Goldberg, M.E., and Bruce C. (1988) Frontal eye field efferents in the macaque monkey. I. Subcortical pathways and topography of striatal and thalamic terminal fields. *J. Comp. Neurol.*, 271: 473–492. DOI: 10.1002/cne.902710402. DOI: 10.1002/cne.902710403. 48

Sylvestre, P.A., and Cullen, K.E. (1999) Quantitative analysis of abducens neuron discharge dynamics during saccadic and slow eye movements. *J. Neurophysiol.*, 82(5), 2612–2632. 32, 90

Van Gisbergen, J.A., Robinson, D.A., and Gielen, S. (1981) A quantitative analysis of generation of saccadic eye movements by burst neurons. *J. Neurophysiol.*, 45(3), 417–442. 16, 20, 27

Van Horn, M.R., Mitchell, D.E., Massot, C., and Cullen, K.E. (2010) Local neural processing and the generation of dynamic motor commands within the saccadic premotor network. *J. Neurosci.*, 30(32): 10905–10917. DOI: 10.1523/JNEUROSCI.0393-10.2010. 40

Walton, M.M.G., Sparks, D.L., and Gandhi N.J. (2005) Simulations of saccade curvature by models that place superior colliculus upstream from the local feedback loop. *J. Neurophysiol.*, 93: 2354–2358. DOI: 10.1152/jn.01199.2004. 90

Wong, W.K., Wang, Z., and Zhen, B. (2012) Relationship between applicability of current based synapses and uniformity of firing patterns. *Int. J. of Neural Systems*, 22(4):1250017. DOI: 10.1142/S0129065712500177. 60, 72

Zajac, F.E. (1989) Muscle and tendon: properties, models, scaling, and application to biomechanics and motor control. *Crit. Rev. Biomed. Eng.*, 17: 359–411. 3

Zee, D.S., Fitzgibbon, E.J., and Optican, L.M. (1992) Saccade-vergence interactions in humans. *J. Neurophysiol.*, 68: 1624–1641. 90

Zhai, X., Ghahari, A., and Enderle, J.D. (2013a) Characteristics of auditory and visual elicited saccades. *Proceedings of the IEEE 39th Northeast Bioengineering Conference*, Syracuse, New York, April 5–7, pp. 183-184. DOI: 10.1109/NEBEC.2013.63.

Zhai, X., Ghahari, A., and Enderle, J.D. (2013b) Parameter estimation of auditory saccades and visual saccades. *Proceedings of the IEEE 39th Northeast Bioengineering Conference*, Syracuse, New York, April 5–7, pp. 185-186. DOI: 10.1109/NEBEC.2013.123. 89

Zhou, W., Chen, X., and Enderle, J.D. (2009) An updated time-optimal 3rd-order linear saccadic eye plant model. *Int. J. of Neural Systems*, Vol. 19, No. 5, 309–330, 2009. DOI: 10.1142/S0129065709002051. 1, 2, 13, 16, 19, 20, 27, 32, 33, 36, 40, 42, 46, 48, 63, 86, 91, 92, 93

Zuber, B.L. (1981) *Models of Oculomotor Behavior and Control* (pp. 21–41). Boca Raton, FL: CRC Press.

Authors' Biographies

ALIREZA GHAHARI

Alireza Ghahari received his B.Sc. degree in electrical engineering from the Sharif University of Technology, Iran, in August 2007. Thereafter, he completed his M.Sc. in electrical and computer engineering at the University of Tehran, Iran, in March 2010. Inspired by the profound contributions of Prof. John Enderle in the field of theoretical and computational neuroscience, he pursued his Ph.D. in electrical and computer engineering with John at the University of Connecticut. After graduation in August 2014, he is currently interested in exploring the field of neural medicine in the context of novel neuroprosthetic methods. Above all, he sees himself as an advocate for the individuals who work to cultivate a sense of gratitude in this contemporary life.

JOHN D. ENDERLE

John D. Enderle is a Professor of Biomedical Engineering and Electrical & Computer Engineering at the University of Connecticut, where he was Biomedical Engineering Program Director from 1997–2010. He received his B.S., M.E., and Ph.D. degrees in biomedical engineering, and a M.E. degree in electrical engineering from Rensselaer Polytechnic Institute, Troy, New York, in 1975, 1977, 1980, and 1978, respectively.

Dr. Enderle is a Fellow of the IEEE, the past Editor-in-Chief of the *EMB Magazine* (2002–2008), the 2004 EMBS Service Award Recipient, Past-President of the IEEE-EMBS, and was EMBS Conference Chair for the 22nd Annual International Conference of the IEEE EMBS and World Congress on Medical Physics and Biomedical Engineering in 2000. He is also a Fellow of the American Institute for Medical and Biological Engineering (AIMBE), Fellow of the American Society for Engineering Education, Fellow of the Biomedical Engineering Society, and a Rensselaer Alumni Association Fellow. Enderle is a former member of the ABET Engineering Accreditation Commission (2004–2009). In 2007, Enderle received the ASEE National Fred Merryfield Design Award. He is also a Teaching Fellow at the University of Connecticut since 1998. Enderle is the Biomedical Engineering Book Series Editor for Morgan & Claypool Publishers.

Enderle is also involved with research to aid persons with disabilities. He is the Editor of the NSF book series on NSF Engineering Senior Design Projects to Aid Persons with Disabilities, published annually since 1989. Enderle is also an author of the book *Introduction to Biomedical Engineering*, published by Elsevier in 2000 (first edition), 2005 (second edition), and 2011 (third edition). Over his career, Enderle has been an author of over 200 publications and 49 books or book chapters. Enderle's current research interest involves characterizing the neurosensory control of the human visual and auditory system.